BeagleBone Cookbook

Mark A. Yoder and Jason Kridner

D1209291

Beijing · Cambridge · Farnham · Köln · Sebastopol · Tokyo

BeagleBone Cookbook

by Mark A. Yoder and Jason Kridner

Copyright © 2015 Mark A. Yoder and Jason Kridner. All rights reserved.

Printed in the United States of America.

Published by O'Reilly Media, Inc., 1005 Gravenstein Highway North, Sebastopol, CA 95472.

O'Reilly books may be purchased for educational, business, or sales promotional use. Online editions are also available for most titles (*http://safaribooksonline.com*). For more information, contact our corporate/institutional sales department: 800-998-9938 or corporate@oreilly.com .

Editors: Brian Sawyer and Rachel Roumeliotis	**Indexer:** Judy McConville
Production Editor: Shiny Kalapurakkel	**Interior Designer:** David Futato
Copyeditor: Bob Russell, Octal Publishing, Inc.	**Cover Designer:** Karen Montgomery
Proofreader: Marta Justak	**Illustrator:** Rebecca Demarest

April 2015:   First Edition

Revision History for the First Edition
2015-03-30:   First Release

See *http://oreilly.com/catalog/errata.csp?isbn=9781491905395* for release details.

The O'Reilly logo is a registered trademark of O'Reilly Media, Inc. *BeagleBone Cookbook*, the cover image, and related trade dress are trademarks of O'Reilly Media, Inc.

978-1-491-90539-5

[LSI]

Table of Contents

Preface

Welcome to *BeagleBone Cookbook*. You are about to cook up some treats with this puppy. BeagleBone Black has put a full-featured web server, Linux desktop, and electronics hub in hundreds of thousands of users' hands, and this book will help you get the most out of it.

Who This Book Is For

This book is intended for users who are either new to BeagleBone Black or are looking to quickly discover some of the many capabilities it provides. Only minimal familiarity with computer programming and electronics is required, because each recipe includes clear and simple wiring diagrams and example code that you can use as a starting point for building experience. Along the way, you'll also find launching-off points to join the large and growing BeagleBoard.org (*http://beagleboard.org/*) community, to further your knowledge and give you a chance to show off what you've learned.

Knowledge of Linux (the operating system kernel upon which the recipes in this book run) is not required. Only a small number of recipes depend on using Linux directly, and we'll try to provide enough background to make you ready to tackle those by the time you get there.

If you don't know what BeagleBone Black is, you might decide to go get one after scanning a few of the recipes. BeagleBone Black is a computer the size of a credit card, yet it is capable of running full desktop applications, such as word processors and web browsers. Unlike other small single-board computers, BeagleBone Black excels at connecting the world of electronics to the world of the Web. The board comes preloaded with a Linux distribution that serves up a web-based integrated development environment (IDE) and all the tools you need to begin creating your own projects without downloading anything or working with tools in the "cloud." Your Bone serves you a programmable cloud all on its own, offering endless possibilities in a variety of fields, including medical, automotive, gaming, software-defined radio, robotics, drones, the Internet of Things (IoT), tools for citizen scientists, and many more. Whatever you develop with the Bone, you have full control of it and the data you choose to share.

This book is also for individuals who care about the future and understanding its direction. With more and more connected technology entering our lives, it is important for us all to have a basic level of understanding of what is possible. Unlike some other small and affordable single-board computers, BeagleBone Black is open hardware. Being a true open-hardware computer, all of the materials for building your own version of the board are freely available on the Web. The design is built on components just about anyone can buy anywhere in the world. Thanks to the open-hardware nature of BeagleBone Black, you can go from the kindergarten of learning about computers and electronics to kickstarting your own business.

How to Use This Book

Although you can read this book cover to cover, each recipe stands alone (aside from a few caveats mentioned in the chapter descriptions in this section), so feel free to browse and jump to the different sections that interest you most. If there's a prerequisite you need to know about, a cross-reference will guide you to the right recipe.

The recipes in the book are organized into the following chapters:

Chapter 1: Basics
    This chapter covers the fundamentals you'll want to know before diving into the recipes in other chapters. We make sure you know how to update the software on your board, create a backup of what you've done, run simple applications, install new applications, and collect some debug information to share if you begin to have some troubles. Looking through the section titles should be enough to know which recipes you can skip.

Chapter 2: Sensors
    This chapter provides recipes on wiring up and collecting data from sensors such as microphones, cameras, buttons, range finders, motion detectors, rotary knobs, and environmental sensors that provide information like humidity, temperature, orientation, position, and much, much more.

Chapter 3: Displays and Other Outputs
    This chapter shows ways to wire up and control displays and other output devices, such as single LEDs, strings of LEDs, LED matrices, LCD displays and external 120 V devices. Motors are saved for Chapter 4. We'll also show you how to make your Bone speak (speech sythesis)

Chapter 4: Motors

In this chapter, you'll learn how to wire up and drive motors of various types, including servo motors that drive to a fixed position, stepper motors that rotate in precise angles, and DC motors that spin quickly. You'll be building mobile robots, 3D printers, CNC mills, and more before you know it. Chapter 8 provides additional coverage of motors, but you'll want to use this chapter first to build up your understanding of motors and how to drive them electrically.

Chapter 5: Beyond the Basics

This chapter offers some additional approaches to solving common challenges a more advanced user might expect or need to know. You'll find coverage of alternative programming languages, environments, and interfaces that many people have found helpful. Fine-tune your Bone and approach to programming it with these tips and tricks.

Chapter 6: Internet of Things

After following the recipes in the first chapters, you have mastered connecting "things" to your Bone. This chapter provides recipes for connecting those things to the Internet for visualizing, controlling remotely, and sharing socially whatever you might desire.

Chapter 7: The Kernel

By this chapter, you should be ready to tackle some more complex challenges, such as updating to a kernel that has support for a bit of hardware not supported in the version you are currently running. We'll even show you how you might add that support to the kernel yourself. Although this chapter isn't for everyone, you'll probably be able to follow the recipes, which will give you a bit more understanding for engaging the Linux experts within the BeagleBoard.org (*http:// beagleboard.org/*) community.

Chapter 8: Real-Time I/O

This chapter provides various solutions for faster input/output (I/O). The solutions range in simplicity from switching to the C programming language, to modifying the kernel to handle real-time requests, to using the programmable real-time unit (PRU).

Chapter 9: Capes

This chapter introduces you to using add-on boards to extend the functionality of BeagleBone Black. We begin by showing you how to use some of the most popular capes and go on to show you how to create your own printed circuit boards (PCBs). Doing so will make the prototypes you create more reliable and easier to assemble, allowing for easier reproduction on a larger number of systems. We'll even show you how to sell hardware you've created over the Internet.

Appendix A: Parts and Suppliers

Throughout the book, we mention various electronic components that you can integrate with your Bone. This appendix summarizes the referenced components, including where to buy them, so you can choose to buy them ahead of time or when you are ready to tackle that given recipe.

Conventions Used in This Book

The following typographical conventions are used in this book:

Italic

Indicates new terms, URLs, email addresses, filenames, and file extensions.

`Constant width`

Used for program listings, as well as within paragraphs to refer to program elements such as variable or function names, databases, data types, environment variables, statements, and keywords.

`Constant width bold`

Shows commands or other text that should be typed literally by the user.

`Constant width italic`

Shows text that should be replaced with user-supplied values or by values determined by context.

This icon signifies a tip or suggestion.

This icon signifies a general note.

This icon indicates a warning or caution.

Using Code Examples

Supplemental material (code examples, exercises, etc.) is available for download at *http://beagleboard.org/cookbook*.

This book is here to help you get your job done. In general, the example code offered with this book may be used in your programs and documentation. You do not need to contact us for permission unless you're reproducing a significant portion of the code. For example, writing a program that uses several chunks of code from this book does not require permission. Selling or distributing a CD-ROM of examples from O'Reilly books does require permission. Answering a question by citing this book and quoting example code does not require permission. Incorporating a significant amount of example code from this book into your product's documentation does require permission.

We appreciate, but do not require, attribution. An attribution usually includes the title, author, publisher, and ISBN. For example: "*BeagleBone Cookbook* by Mark A. Yoder (Rose-Hulman Institute of Technology) and Jason Kridner (Texas Instruments and BeagleBoard.org Foundation). Copyright 2015 Mark A. Yoder and Jason Kridner, 978-1491905395."

If you feel your use of code examples falls outside fair use or the permission given above, feel free to contact us at *permissions@oreilly.com*.

About the Diagrams

Many of the projects presented in this book include diagrams documenting the wiring required to connect hardware to the BeagleBone. Our diagrams are created with Fritzing (*http://fritzing.org/home/*). You can install Fritzing on your host computer by following the download instructions (*http://fritzing.org/download/*). Adafruit has a nice library of parts, which includes the Bone. These instructions (*http://elinux.org/Beagleboard:Fritzing_on_the_BeagleBone_Black*) tell you how to load it.

After you have everything installed, head to the Fritzing Tutorial (*http://fritzing.org/learning/tutorials*) page to learn how to use it.

Fritzing is a useful open source program for creating hardware illustrations. With it, you can create breadboard diagrams, schematic diagrams, and designs for PCBs. Figure P-1 shows the basic setup used for all the diagrams in this book.

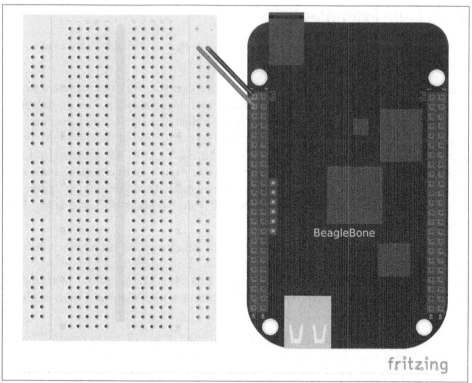

Figure P-1. Basic Fritzing setup

We've used a half-sized breadboard, because it's about the same size as the Bone.

Safari® Books Online

 *Safari Books Online* is an on-demand digital library that delivers expert content in both book and video form from the world's leading authors in technology and business.

Technology professionals, software developers, web designers, and business and creative professionals use Safari Books Online as their primary resource for research, problem solving, learning, and certification training.

Safari Books Online offers a range of plans and pricing for enterprise, government, education, and individuals.

Members have access to thousands of books, training videos, and prepublication manuscripts in one fully searchable database from publishers like O'Reilly Media, Prentice Hall Professional, Addison-Wesley Professional, Microsoft Press, Sams, Que,

Peachpit Press, Focal Press, Cisco Press, John Wiley & Sons, Syngress, Morgan Kaufmann, IBM Redbooks, Packt, Adobe Press, FT Press, Apress, Manning, New Riders, McGraw-Hill, Jones & Bartlett, Course Technology, and hundreds more. For more information about Safari Books Online, please visit us online.

How to Contact Us

Please address comments and questions concerning this book to the publisher:

O'Reilly Media, Inc.
1005 Gravenstein Highway North
Sebastopol, CA 95472
800-998-9938 (in the United States or Canada)
707-829-0515 (international or local)
707-829-0104 (fax)

We have a web page for this book, where we list errata, examples, and any additional information. You can access this page at *http://www.oreilly.com/catalog/0636920033899.do*.

To comment or ask technical questions about this book, send email to *bookquestions@oreilly.com*.

For more information about our books, courses, conferences, and news, see our website at *http://www.oreilly.com*.

Find us on Facebook: *http://facebook.com/oreilly*

Follow us on Twitter: *http://twitter.com/oreillymedia*

Watch us on YouTube: *http://www.youtube.com/oreillymedia*

Acknowledgments

We thank Fred Schroeder, Charles Steinkuhler, Christine Long, Martee Held, David Scheltema, and Drew Fustini for their immensely helpful reviews on both content and approach. We also thank our editor, Brian Sawyer, for his work in leading us through this process and keeping us focused.

## CHAPTER 1

Basics

1.0 Introduction

When you buy BeagleBone Black, pretty much everything you need to get going comes with it. You can just plug it into the USB of a host computer, and it works. The goal of this chapter is to show what you can do with your Bone, right out of the box. It has enough information to carry through the next three chapters on sensors (Chapter 2), displays (Chapter 3), and motors (Chapter 4).

1.1 Picking Your Beagle

Problem

There are five different BeagleBoards. How do you pick which one to use?

Solution

For someone new to the BeagleBoard.org boards, the BeagleBone Black is the obvious choice. It's the newest and cheapest (~$55 USD) of the inexpensive Bones, and there are nearly a quarter-million units out there. Many of the recipes in this book will work on the other Beagles, too, but the Black is where to begin. If you already have BeagleBone Black, you can move on to the next recipe. If you have another Beagle, your mileage may vary.

Discussion

If you already have a BeagleBoard and it isn't a Black, read on to see what it can do.

There is a high-end line and a low-end line for BeagleBoard.org boards. The Beagle*Boards* are the higher-end boards, with strongest appeal to advanced Linux hackers looking for affordable solutions with digital signal processors (DSPs) and features that can improve display performance. The Beagle*Bones* are the lower-end boards, with strongest appeal to everyday "makers" of things, including lots of low-level I/O, but they still have reasonable display performance.

The Original BeagleBoard

The original BeagleBoard (Figure 1-1) came out in July 2008.

Figure 1-1. The original BeagleBoard

The original BeagleBoard used the very first ARM Cortex-A8 processor, starting at 600 MHz with 256 MB of RAM. The clock frequency was then increased to 720 MHz. Figure 1-2 shows more details of its original features.

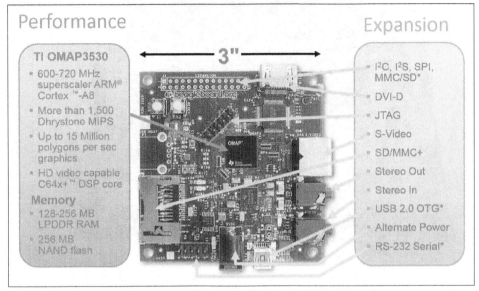

Figure 1-2. Details of the original BeagleBoard

There are many in the field, and they are still being sold, but the newer Bones are a better starting point.

BeagleBoard-xM

The BeagleBoard-xM (Figure 1-3) began shipping August 2010.

Figure 1-3. The BeagleBoard-xM

The xM brings with it a faster ARM processor running at 1 GHz, twice the RAM as the original at 512 MB, and four USB ports. Figure 1-4 shows additional features. The xM sells for $150 USD.

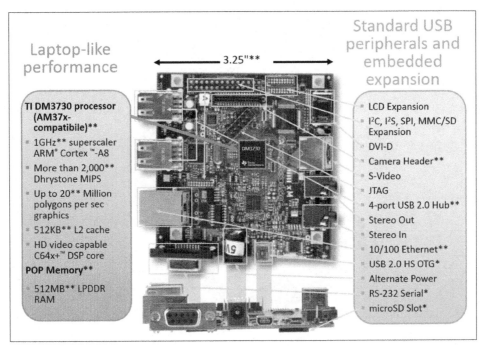

Figure 1-4. Details of the BeagleBoard-xM

The xM is a good choice if you want to use the BeagleBoard as a small workstation. You can plug in a USB keyboard, mouse, and an HDMI display. Add a 5 V power supply, and you are ready to run without the need of a host computer.

BeagleBone White

The original BeagleBone appeared in October 2011. Shown in Figure 1-5, it's white (thus the name BeagleBone White), and it sells for $90 USD. It dropped back to 256 MB RAM and 720 MHz clock, but it can fit in an Altoids tin.

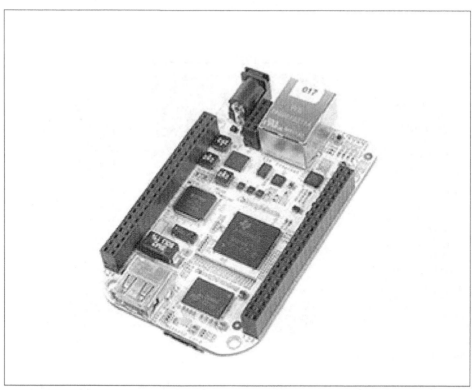

Figure 1-5. The original BeagleBone, known as the White

The BeagleBone also adds two 46-pin expansion headers (shown in Figure 1-6) and the ability to add capes (Chapter 9).

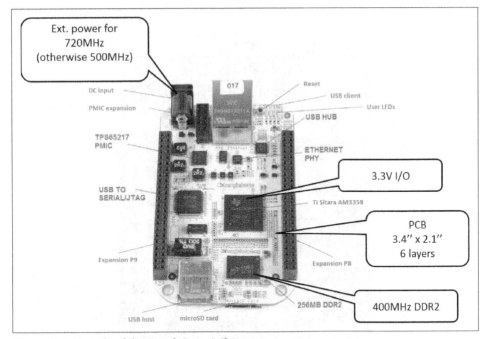

Figure 1-6. Details of the BeagleBone White

The White can run the same software images off the same SD cards as the Black does, so everything presented in this book will also run on the White.

BeagleBone Black

April 2013 saw the introduction of BeagleBone Black (shown in Figure 1-7), which is the real focus of the recipes in this book.

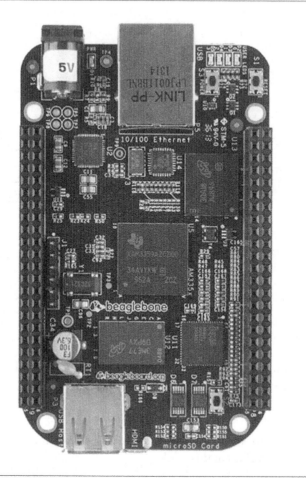

Figure 1-7. The BeagleBone Black

The Black initially reduced the price by half, to $45 USD, bumped the RAM back to 512 MB, and pushed the clock to 1 GHz. It also added 2 GB of onboard flash memory. Rev C of the board increased the onboard flash to 4 GB and raised the price to $55 USD. All this, and, again, it still fits in an Altoids tin.

Figure 1-8 provides more details on the BeagleBone Black.

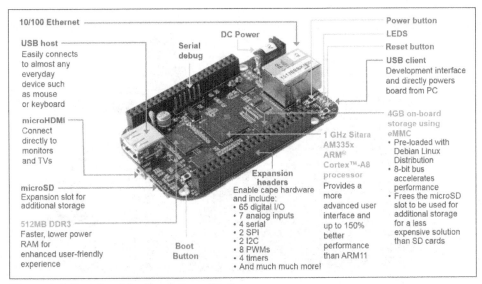

Figure 1-8. Details of the BeagleBone Black

This book focuses on BeagleBone Black (simply called the Bone here), so more details will follow.

BeagleBoard-X15

At the time of this writing (late 2014/early 2015), the newest addition to the orginal BeagleBoard lineup has not yet had any official details announced. However, due to the open development nature of BeagleBoard.org, many *potential* details of BeagleBoard-X15 have been discussed in public mailing lists and blog posts, such as the post on the cnx-software blog (*http://bit.ly/1xcS1y6*). The concept is to have support in the mainline u-boot and Linux kernel projects ahead of public hardware availability. The BeagleBoard wiki (*http://bit.ly/1B4pStp*) provides the current state of development. Figure 1-9 shows an image of the current prerelease development board.

Figure 1-9. The BeagleBoard-X15

The current prerelease board measures 4″ x 4.2″, is based on a Texas Instruments dual-core ARM Cortex-A15 processor running at 1.5 GHZ, and features 2 GB of DDR3L memory. It includes an estimated 157 general-purpose input/output (GPIO) pins and interfaces for audio I/O, simultaneous LCD and HDMI video output, PCIe, eSATA, USB3, and dual-gigabit Ethernet. Additional processing elements include dual TI C66x DSPs, dual ARM Cortex-M4s, dual PRU-ICSS subsystems, and multiple image/graphics processing engines.

BeagleBoard-X15 is going to be an amazing machine, but it will be a fair bit more expensive and complicated. Starting with BeagleBone Black and moving up to BeagleBoard-X15 when you've identified a clear need for its power is a good approach.

1.2 Getting Started, Out of the Box

Problem

You just got your Bone, and you want to know what to do with it.

Solution

Fortunately, you have all you need to get running: your Bone and a USB cable. Plug the USB cable into your host computer (Mac, Windows, or Linux) and plug the mini-USB connector side into the USB connector near the Ethernet connector on the Bone, as shown in Figure 1-10.

Figure 1-10. Plugging BeagleBone Black into a USB port

The four blue USER LEDs will begin to blink, and in 10 or 15 seconds, you'll see a new USB drive appear on your host computer. Figure 1-11 shows how it will appear on a Windows host, and Linux and Mac hosts will look similar. The Bone acting like a USB drive and the files you see are located on the Bone.

Figure 1-11. The Bone appears as a USB drive

Open the drive and open *START.htm* using Google Chrome (*http://www.google.com/chrome/*) or Firefox (*https://www.mozilla.org/en-US/firefox/new/*) (Figure 1-12).

 Some users have reported problems when using Internet Explorer with the web pages served up by the Bone, so make sure to use Chrome or Firefox.

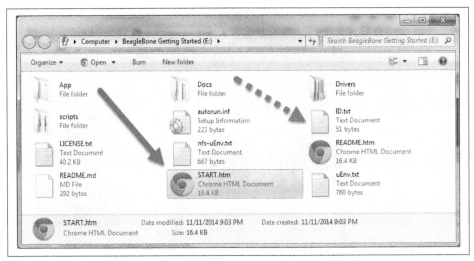

Figure 1-12. Open START.htm

Follow the instructions (Figure 1-13) for installing the USB drivers for your host computer's operating system (OS).

Figure 1-13. Install the USB drivers

On your host, browse to *http://192.168.7.2* (Figure 1-14).

Figure 1-14. Open http://192.168.7.2 on your host computer

You are now ready to explore your Bone. Look around. There's lots of information on the page.

 The green banner at the top of the page means the browser is talking to the Bone, and code on the page can be edited and run on the Bone. Try scrolling down to the code in "BoneScript interactive guide" and running it. Then try editing the code and running it again. Take five minutes and try it! Watch out, though, because you can't save your edits. You need Cloud9 for that, as discussed next.

Also be sure to browse to *http://192.168.7.2:3000* from your host computer (Figure 1-15).

Figure 1-15. Cloud9

Here, you'll find *Cloud9*, a web-based integrated development environment (IDE) that lets you edit and run code on your Bone! See Recipe 1.6 for more details.

 Cloud9 can have different themes. If you see a white background, you can match the cookbook's figures by clicking on the *Main Theme* drop-down menu and selecting *Cloud9 Classic Dark Theme*.

 Make sure you turn off your Bone properly. It's best to run the `halt` command:

```
bone# halt
The system is going down for system halt NOW! (pts/0)
```

This will ensure that the Bone shuts down correctly. If you just pull the power, it's possible that open files won't close properly and might become corrupt.

Discussion

The rest of this book goes into the details behind this quick out-of-the-box demo. Explore your Bone and then start exploring the book.

1.3 Verifying You Have the Latest Version of the OS on Your Bone

Problem

You just got BeagleBone Black, and you want to know which version of the operating system it's running.

Solution

This book uses Debian (*https://www.debian.org*), the Linux distribution that currently ships on the Bone. However this book is based on a newer version (2014-11-11 image) than what is shipping at the time of this writing. You can see which version your Bone is running by following the instructions in Recipe 1.2 to open the USB drive that comes from the Bone, as shown in Figure 1-12. But instead of opening *START.html*, open *ID.txt* (shown with the dashed arrow in Figure 1-12). You'll see something like Figure 1-16, in which 2014-11-11 is the date of the image.

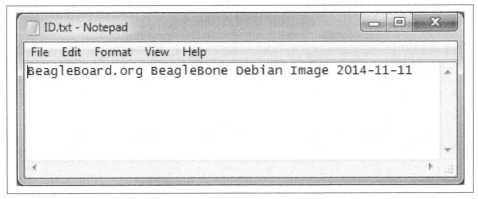

Figure 1-16. Contents of ID.txt

Discussion

If you don't find *ID.txt* on your drive, or if the date given isn't 2014-11-11 or newer, you *may* need to update your image, as described in Recipe 1.10.

Many of the following examples in this chapter and Chapter 2, Chapter 3, and Chapter 4 may run fine on an older image. Try it and see. If you run into problems, see Recipe 1.10 to update your Bone's OS version.

1.4 Running the BoneScript API Tutorials

Problem

You'd like to learn JavaScript and the BoneScript API to perform physical computing tasks without first learning Linux.

Solution

Plug your board into the USB of your host computer and browse to *http://192.168.7.2/Support/bone101/* using Google Chrome or Firefox (as shown in Recipe 1.2). In the left column, click the BoneScript title, which will take you to *http://192.168.7.2/Support/BoneScript/* (Figure 1-17).

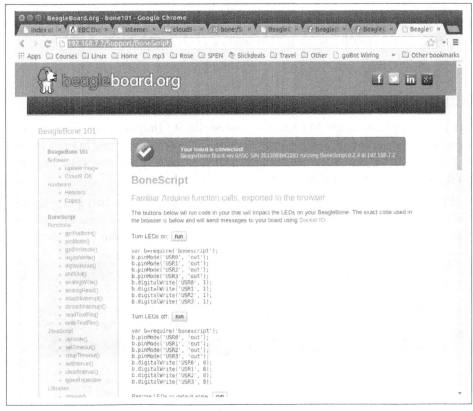

Figure 1-17. The BoneScript API examples page

Explore the various demonstrations of the BoneScript API. The BoneScript examples page (*http://192.168.7.2/Support/BoneScript*) lists several places to learn more about JavaScript and BoneScript (Figure 1-17).

If the banner is green, the examples are live. Clicking the "run" button will make them run on your Bone.

Here's yet another place to explore. In the left column of Figure 1-17, click the function names. Take five minutes and see what you can find.

You can edit the JavaScript on the BoneScript API examples page, but you can't save it for later. If you want to edit and save it for later, fire up Cloud9 (Recipe 1.6) and look in the *examples* folder.

Discussion

The JavaScript BoneScript API tutorials use Socket.IO (*http://socket.io*) to create a connection between your browser and your Bone. The editor running in these tutorials gives you the ability to edit the examples quickly and try out new code snippets, but it doesn't provide the ability to save.

Visit the BONE101 GitHub page (*http://beagleboard.github.io/bone101*) and select the "Fork me on GitHub" link (Figure 1-18) to learn more about how these tutorials are implemented and how you can embed similar functionality into your own web pages.

Figure 1-18. Bone101 GitHub page

1.5 Wiring a Breadboard

Problem

You would like to use a breadboard to wire things to the Bone.

Solution

Many of the projects in this book involve interfacing things to the Bone. Some plug in directly, like the USB port. Others need to be wired. If it's simple, you might be able to plug the wires directly into the P8 or P9 headers. Nevertheless, many require a breadboard for the fastest and simplest wiring.

To make this recipe, you will need:

- Breadboard and jumper wires (see "Prototyping Equipment" on page 316)

Figure 1-19 shows a breadboard wired to the Bone. All the diagrams in this book assume that the ground pin (P9_1 on the Bone) is wired to the negative rail and 3.3 V (P9_3) is wired to the positive rail.

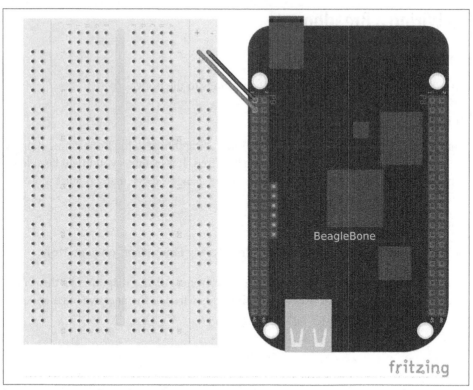

Figure 1-19. Breadboard wired to BeagleBone Black

Discussion

There are many good tutorials online about wiring breadboards. (See How to Use a Breadboard [*http://bit.ly/1NJMFom*] and Breadboard Basics for Absolute Beginners [*http://bit.ly/1FaPB9M/*], for example). The key thing to know is which holes are attached to which. Figure 1-20 gives a clear illustration of the connections.

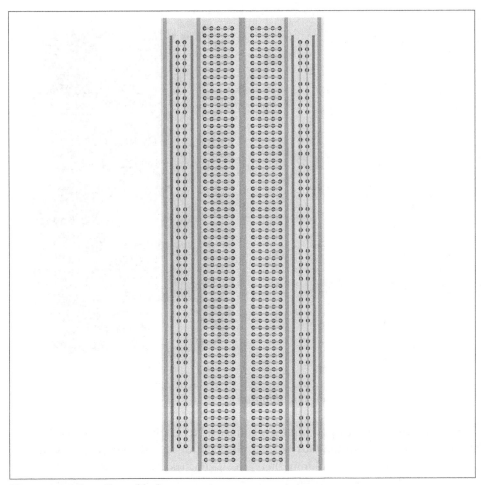

Figure 1-20. Breadboard hole connections

The *rails*, or *buses*, are the pairs of columns of holes that run up and down on the left and right side of the board. All the holes in one column are electrically wired together. Therefore, attaching ground to the rightmost column makes ground available everywhere on that column. Likewise with attaching 3.3 V to the next column over, it's easily wired from anywhere up and down that column.

In the middle, groups of five holes are wired from left to right, as shown in Figure 1-20. For example, Figure 1-21 shows a pushbutton switch plugged into four holes.

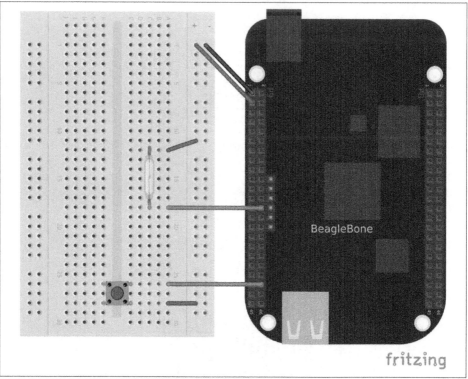

Figure 1-21. Diagram for wiring a pushbutton

The lower-right pin of the pushbutton is connected to a hole of the breadboard that is electrically connected to the four holes to the right of it. The rightmost hole of that group is wired to the plus column; therefore, it attaches the pushbutton to the 3.3 V on the Bone. The upper-right pin of the pushbutton is electrically connected to the four holes to the right of it; the rightmost pin is wired to GPIO pin P9_42 on the Bone.

If you're careful when wiring to keep the wires flat on the board, it's easy to understand and debug a moderately complex circuit.

 By default, many of the Bone's GPIO pins are configured as inputs, such as we would want in this circuit. Later, we'll show you that they can be configured as outputs, and thus you should take care to avoid shorting an output pin. If the switches were connected as shown in Figure 1-21 and the GPIO pins were configured as outputs, having the buttons active would cause the GPIO pin to be shorted to 3.3 V. If the pin were outputting a low or 0, it would try to sink all the current it could to make the pin go low. The result is that the pin would burn out.

In addition to the GPIO pins being able to be either inputs or outputs, there are also internal pull-up and pull-down resistors on the Bone's GPIO pins. It is necessary to use the pull-down mode of the GPIO pins for this circuit to work predictably. Otherwise, when the switches are open, the input pins are not connected electrically and will be in an unknown state.

1.6 Editing Code Using the Cloud9 IDE

Problem

You want to edit and debug files on the Bone.

Solution

Plug your Bone into a host computer via the USB cable. Open a browser (either Google Chrome or FireFox will work) on your host computer (as shown in Recipe 1.2). After the Bone has booted up, browse to *http://192.168.7.2:3000* on your host. You will see something like Figure 1-15.

Click the *examples* folder on the left and then double-click *blinkled.js*. You can now edit the file. If you would like to edit files in your home directory, on the left of the Cloud9 screen, go to the Preferences wheel in the Workspace browser and select Add Home to Favorites (Figure 1-22). Now, your home directory will show up in the Workspace file tree.

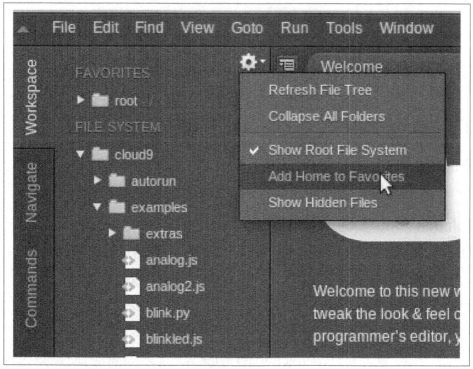

Figure 1-22. Making your home folder appear in Cloud9

If you edit line 13 of the *blinkled.js* file (setInterval(toggle, 1000);), changing 1000 to 500, you must save the file before running it for the change to take effect. The blue LED next to the Ethernet port on your Bone will flash roughly twice as fast.

Figure 1-22 shows */root* has been added under *FAVORITES*.

The *cloud9* folder that appears under *FILE SYSTEM* is located in */var/lib/cloud9* on your Bone.

Discussion

The Bone comes with a nice web-based IDE called *Cloud9*. You can learn all about it at the Cloud9 website (*https://c9.io/*), although it's easy enough to use that you can just dive in and begin using it.

By default, the Bone's operating system runs the Cloud9 IDE as the root user. In Linux, this is a special account that has privileges to perform many operations often excluded to normal user accounts. These operations include accessing various hardware features and contents of various files, along with the ability to modify the on-board flash contents. This was done to make many tasks simpler to complete, but caution should be used before editing system files you don't understand. The good news is that you can always read Recipe 1.13 to learn how to restore the onboard flash to the factory contents.

You can find out more about the root account on the Linux Documentation Project System Administrator's guide's entry on accounts (*http://bit.ly/1C575DD*) and Debian's wiki entry on the root user (*https://wiki.debian.org/Root*).

1.7 Running JavaScript Applications from the Cloud9 IDE

Problem

You have a file edited in Cloud9, and you want to run it.

Solution

Cloud9 has a `bash` command window built in at the bottom of the window. You can run your code from this window. To do so, add `#!/usr/bin/env node` at the top of the file that you want to run and save.

If you are running Python, replace the word `node` in the line with `python`.

At the bottom of the Cloud9 window are a series of tabs (Figure 1-23). Click the `bash` tab (it should be the leftmost tab). Here, you have a command prompt. In my case, it's `root@yoder-debian-bone:/var/lib/cloud9#`. Yours will be slighly different, but it should end with a #.

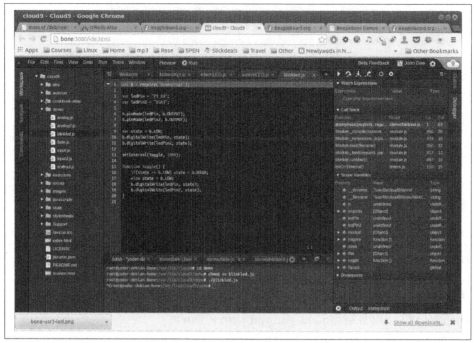

Figure 1-23. Cloud9 debugger

Change to the directory that contains your file, make it executable, and then run it:

```
root@bone:/var/lib/cloud9# cd examples
root@bone:/var/lib/cloud9/examples# chmod +x blinkled.js
root@bone:/var/lib/cloud9/examples# ./blinkled.js
```

The cd is the change directory command. After you cd, you are in a new directory, and the prompt reflects that change. The chmod command changes the mode of the file. The +x indicates that you want to add execute permission. You need to use the chmod +x command only once. Finally, ./blinkled.js instructs the JavaScript to run. You will need to press ^C (Ctrl-C) to stop your program.

Discussion

The Debian distribution we are using comes with two users: root and debian. The bash command window for Cloud9 runs as root; you can tell this by looking at the last character of the prompt. If it's a #, as shown here, you are running as root. If it's a $, you are running as a normal user (debian).

In this book, we assume that you are running as `root`.

`root` is also known as the superuser. Running as `root`, you can access most everything on the Bone. This is both good and bad. If you know what you are doing, you can easily interact with all of the hardware. But, if you don't know what you are doing, you can really mess things up. If you follow our examples carefully, you shouldn't have any problems.

Initially, the `root` password isn't set, so anyone can log in without a password. This may be OK if your Bone is never connected to the network, but it's better practice to set a password:

```
bone# passwd
Enter new UNIX password:
Retype new UNIX password:
passwd: password updated successfully
```

For security, the password you type won't be displayed.

1.8 Running Applications Automatically

Problem

You have a BoneScript application that you would like to run every time the Bone starts.

Solution

This is an easy one. In Cloud9, notice the folder called *autorun* (Figure 1-24). Place any BoneScript files you want to run at boot time in this folder. The script will begin execution immediately and will stop execution when you remove the file from this folder.

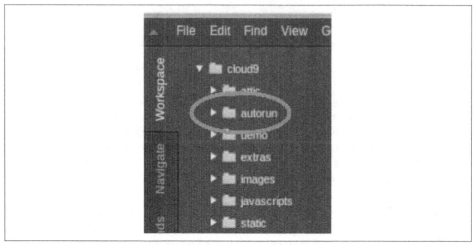

Figure 1-24. Making applications autorun at boot-up time

You can drag and drop the script into the *autorun* folder using the Cloud9 IDE workspace view, or you can move it using the bash prompt:

```
bone# mv myApp.js autorun
```

Discussion

Your script will start running as soon as you copy it to the folder and will run every time you boot up the Bone. In fact, it will run as soon as you put it in the folder, and if you change the file, it will restart with the current contents.

This solution currently works with Node.js applications, Python scripts, and *.ino* Arduino sketches (files ending in *.js*, *.py*, or *.ino*, respectively). You can check the BoneScript GitHub repository (*http://bit.ly/1EbB9fW*) to see if any other types have been added.

1.9 Finding the Latest Version of the OS for Your Bone

Problem

You want to find out the latest version of Debian that is available for your Bone.

Solution

At the time they were written, these instructions were up-to-date. Go to *http://beagleboard.org/latest-images* for the latest instructions.

On your host computer, open a browser and go to *http://rcn-ee.net/deb/testing/*. This shows you a list of dates of the most recent Debian images (Figure 1-25).

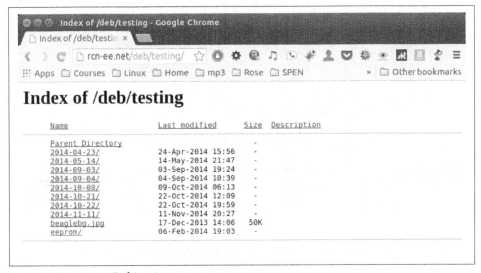

Figure 1-25. Latest Debian images

Clicking a date will show you several variations for that particular date. Figure 1-26 shows the results of clicking *2014-11-11*.

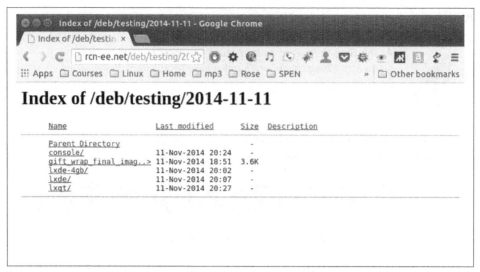

Figure 1-26. Latest Debian images for a given date

Clicking *lxde-4gb/* shows a list of 4 GB images (Figure 1-27).

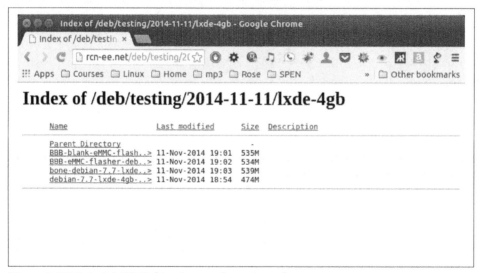

Figure 1-27. Latest 4 GB Debian images for a given date

These are the images you want to use if you are flashing a Rev C BeagleBone Black onboard flash, or flashing a 4 GB or bigger miscroSD card. The image beginning with *bone-debian-7.7-lxde* is used for programming the microSD card. The one beginning with *BBB-eMMC-flasher-deb* is for programming the onboard flash memory.

The onboard flash is often called the *eMMC* memory. We just call it *onboard flash*, but you'll often see *eMMC* appearing in filenames of images used to update the onboard flash.

Discussion

There are a number of images posted on this site. The subdirectory names, such as *lxde-4gb*, might change with time and might take some investigating to learn what they mean. In this case, a simple Google search reveals that *lxde* is the Lightweight X11 Desktop Environment (*http://lxde.org/*). The *-4gb* means it's a 4 GB image. The other *lxde* has 2 GB images, which are needed for the pre-Rev C BeagleBone Blacks, because their onboard flash is only 2 GB. The *lxqt* has images that use the next generation of the Lightweight Desktop Environment (*http://lxqt.org/*). These images are for a newer release of Debian.

You can try running a newer release, but things might not work exactly as described in this text. See Recipe 5.15 for information about the Debian release system.

The images suggested here are big installations with many important applications already included. If you want an image with almost nothing included, try *console* (Figure 1-26). The images in this folder are very minimal, with very few programs installed (for example, Cloud9 and BoneScript are missing), but they have enough to boot up and run. Notice that they are about one-tenth the size of the complete installs.

1.10 Running the Latest Version of the OS on Your Bone

Problem

You want to run the latest version of the operating system on your Bone without changing the onboard flash.

Solution

This solution is to flash an external microSD card and run the Bone from it. If you boot the Bone with a microSD card inserted with a valid boot image, it will boot from the microSD card. If you boot without the microSD card installed, it will boot from the onboard flash.

If you want to reflash the onboard flash memory, see Recipe 1.13.

 I instruct my students to use the microSD for booting. I suggest they keep an extra microSD flashed with the current OS. If they mess up the one on the Bone, it takes only a moment to swap in the extra microSD, boot up, and continue running. If they are running off the onboard flash, it will take much longer to reflash and boot from it.

Windows

If you are using a host computer running Windows, go to *http://rcn-ee.net/deb/test ing/2014-11-11/lxde-4gb/*, and download *bone-debian-7.7-lxde-4gb- armhf-2014-11-11-4gb.img.xz*. It's more than 500 MB, so be sure to have a fast Internet connection. Then go to *http://beagleboard.org/getting-started#update* and follow the instructions there to install the image you downloaded.

Linux

If you are running a Linux host, plug a 4 GB byte or bigger microSD card into a reader on your host and run Disks.

Select the microSD Drive and unmount (Figure 1-28) any partitions that have mounted. Note the path to the device (shown with an arrow in Figure 1-28) at the top of the Disks window. In my case, it's */dev/sdb*. We'll use this path in a moment.

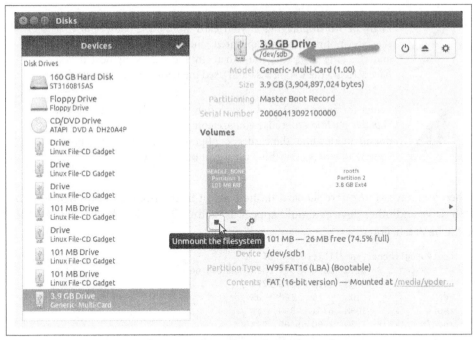

Figure 1-28. Unmounting the microSD card via the Disks application

Run the following command to download the 2014-11-11 image (be sure that you have a fast Internet connection; it's more than 500 MB in size):

```
host$ wget http://rcn-ee.net/deb/testing/2014-11-11/lxde-4gb/\
bone-debian-7.7-lxde-4gb-armhf-2014-11-11-4gb.img.xz
```

This will copy the disk image to the current directory the command was run from to your host computer. This will take a couple minutes on a fast connection.

The downloaded file is compressed. Uncompress it by using the following command:

```
host$ unxz bone-debian-7.7-lxde-4gb-armhf-2014-11-11-4gb.img.xz
```

After a minute or so, the compressed *.imgxz* file will be replaced by the uncompressed *.img* file. Then write it to the microSD card by using the following command, substituting your device path noted earlier (/dev/sdb, in my case) for the device path given in the dd command:

```
host$ sudo dd if=bone-debian-7.7-lxde-4gb-armhf-2014-11-11-4gb.img \
        of=/dev/sdb bs=8M
```

The dd command takes 5 to 10 minutes.

This operation will wipe out everything on the microSD card. It might be worth plugging in your card, noting the path, removing the card, noting it has disappeared, and then plugging it in again and checking the path. You can really mess up your host if you have selected the wrong disk and used the wrong path. Be careful.

When formatting SD cards, you often need to be sure to have a *bootable partition*. Because you are completly rewriting the card, it doesn't matter how the card is configured before writing. The dd command writes everything the way it needs to be.

When you have your microSD card flashed, put it in the Bone and power it up. The USB drive and other devices should appear as before. Open Cloud9 (Recipe 1.6) and, in the bash tab, enter:

```
root@beaglebone:/var/lib/cloud9# df -h
Filesystem       Size  Used Avail Use% Mounted on
rootfs           3.2G  2.0G  1.0G  29% /
udev              10M     0   10M   0% /dev
tmpfs            100M  676K   99M   1% /run
/dev/mmcblk0p2   7.2G  2.0G  5.0G  29% /
tmpfs            249M     0  249M   0% /dev/shm
tmpfs            249M     0  249M   0% /sys/fs/cgroup
tmpfs            5.0M     0  5.0M   0% /run/lock
tmpfs            100M     0  100M   0% /run/user
/dev/mmcblk0p1    96M   62M   35M  65% /media/BEAGLEBONE
/dev/mmcblk1p2   1.8G  290M  1.4G  18% /media/rootfs
/dev/mmcblk1p1    16M  520K   16M   4% /media/BEAGLEBONE_
```

This prints out how much of the disk is free. The first line is the one we're interested in. If the Size is much smaller than the size of your microSD card, you'll need to resize your partition. Just enter the following:

```
root@beaglebone:/var/lib/cloud9# cd /opt/scripts/tools/
root@beaglebone:/opt/scripts/tools# ./grow_partition.sh
root@beaglebone:/opt/scripts/tools# reboot
root@beaglebone:/var/lib/cloud9# df -h
Filesystem       Size  Used Avail Use% Mounted on
rootfs           7.2G  2.0G  5.0G  29% /
udev              10M     0   10M   0% /dev
tmpfs            100M  676K   99M   1% /run
/dev/mmcblk0p2   7.2G  2.0G  5.0G  29% /
tmpfs            249M     0  249M   0% /dev/shm
tmpfs            249M     0  249M   0% /sys/fs/cgroup
tmpfs            5.0M     0  5.0M   0% /run/lock
tmpfs            100M     0  100M   0% /run/user
/dev/mmcblk0p1    96M   62M   35M  65% /media/BEAGLEBONE
/dev/mmcblk1p2   1.8G  290M  1.4G  18% /media/rootfs
/dev/mmcblk1p1    16M  520K   16M   4% /media/BEAGLEBONE_
```

This clever script will figure out how big the partition can be and grow it to that size. A reboot is necessary.

Here, I started by putting a 4 GB image on an 8 GB microSD card. Initially, only 3.2 GB were usable. After growing the partition, most of the card (7.2 GB) is available.

Mac

If you are running from a Mac host, the steps are fairly similar to running on a Linux host, except that you won't be able to view the Linux partition on the created microSD card.

Begin by plugging a 4 GB or bigger microSD card into a reader on your host and then run Disk Utility. Select the disk and click Info. In Figure 1-29, you can see the Disk Identifier is disk1s1.

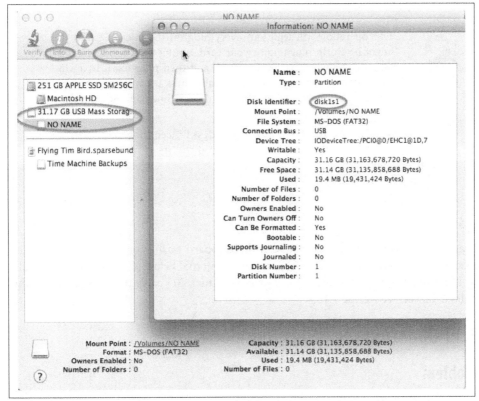

Figure 1-29. Examining the microSD card via the Disk Utility application

The important part of the Disk Identifier is the number immediately following disk (a 1 in Figure 1-29). We'll use this identifier to overwrite the microSD contents.

From your Mac's Terminal, run the following command to download the *2014-11-11* image (again, be sure that you have a fast Internet connection, because it's more than 500 MB):

```
mac$ curl -O http://rcn-ee.net/deb/testing/2014-11-11/lxde-4gb/\
bone-debian-7.7-lxde-4gb-armhf-2014-11-11-4gb.img.xz
```

You'll need to have the xz utility installed (download from The Tukaani Project [*http://tukaani.org/xz/*]). Uncompress the image by using the following command (this will take a minute or so):

```
mac$ unxz bone-debian-7.7-lxde-4gb-armhf-2014-11-11-4gb.img.xz
```

Then write it to the microSD card, substituting your device path noted earlier (/dev/ rdisk1, in my case) for the device path given in the dd command:

```
mac$ sudo dd if=bone-debian-7.7-lxde-4gb-armhf-2014-11-11-4gb.img of=/dev/rdisk1
```

You'll need to type in your password. The dd command takes 5 to 10 minutes.

This operation will wipe out everything on the microSD card. It might be worth plugging in your card, noting the path, removing the card, noting it has disappeared, and then plugging it in again and checking the path. You can really mess up your host if you have selected the wrong disk and used the wrong path. Be careful.

Note that I used rdisk1 rather than disk1. According to the eLinux wiki (*http://bit.ly/1BqOxwW*), doing so will speed up writing quite a bit.

Discussion

BeagleBone Black can run its OS from the onboard flash memory (2 GB on older Blacks and 4 GB on the new Rev Cs) or from a microSD card. The 2 GB is a bit small, so I advise my students to use a 4 GB or bigger microSD card. If you want to use the onboard flash, see Recipe 1.13.

1.11 Updating the OS on Your Bone

Problem

You've installed the latest version of Debian on your Bone (Recipe 1.10), and you want to be sure it's up-to-date.

Solution

Ensure that your Bone is on the network and then run the following command on the Bone:

```
bone# apt-get update
bone# apt-get upgrade
```

If there are any new updates, they will be installed.

 If you get the error The following signatures were invalid: KEYEXPIRED 1418840246, see eLinux support page (*http://bit.ly/1EXocb6*) for advice on how to fix it.

Discussion

After you have a current image running on the Bone, it's not at all difficult to keep it upgraded.

1.12 Backing Up the Onboard Flash

Problem

You've modified the state of your Bone in a way that you'd like to preserve or share.

Solution

The eLinux (*http://elinux.org/*) page on BeagleBone Black Extracting eMMC contents (*http://bit.ly/1C57I0a*) provides some simple steps for copying the contents of the onboard flash to a file on a microSD card:

1. Get a 4 GB or larger microSD card that is FAT formatted.

2. If you create a FAT-formatted microSD card, you must edit the partition and ensure that it is a bootable partition.

3. Download beagleboneblack-save-emmc.zip (*http://bit.ly/1wtXwNP*) and uncompress and copy the contents onto your microSD card.

4. Eject the microSD card from your computer, insert it into the powered-off BeagleBone Black, and apply power to your board.

5. You'll notice USER0 (the LED closest to the S1 button in the corner) will (after about 20 seconds) begin to blink steadily, rather than the double-pulse "heartbeat" pattern that is typical when your BeagleBone Black is running the standard Linux kernel configuration.

6. It will run for a bit under 10 minutes and then USER0 will stay on steady. That's your cue to remove power, remove the microSD card, and put it back into your computer.

7. You will see a file called *BeagleBoneBlack-eMMC-image-XXXXX.img*, where *XXXXX* is a set of random numbers. Save this file to use for restoring your image later.

 Because the date won't be set on your board, you might want to adjust the date on the file to remember when you made it. For storage on your computer, these images will typically compress very well, so use your favorite compression tool.

 The eLinux wiki (*http://elinux.org/Beagleboard*) is the definitive place for the BeagleBoard.org community to share information about the Beagles. Spend some time looking around for other helpful information.

Discussion

The *beagleboneblack-save-emmc.zip* file contains just enough of a Linux operating system to know how to copy an image from the onboard flash to a file. The code that does the work is in *autorun.sh*, which is shown in Example 1-1.

Example 1-1. Code for copying from onboard flash to microSD card

```
#!/bin/sh
echo timer > /sys/class/leds/beaglebone\:green\:usr0/trigger
dd if=/dev/mmcblk1 of=/mnt/BeagleBoneBlack-eMMC-image-$RANDOM.img bs=10M
sync
echo default-on > /sys/class/leds/beaglebone\:green\:usr0/trigger
```

The first echo changes the USER0 LED to flash a different pattern to show the copy is being made. The dd command then copies the onboard flash image from input file (if) */dev/mmcblk1* to an output file (of) on the microSD card called */mnt/ BeagleBoneBlack-eMMC-image-$RANDOM.img*. */dev/mmcblk1* is where the entire onboard flash appears as a raw image and dd just copies it.

The final echo turns the USER0 LED on to notify you that the copying is over.

You can use the same microSD card to copy an *.img* file back to the onboard flash. Just edit the *autorun.sh* file to include Example 1-2. This code is largely the same as Example 1-1, except it reverses the direction of the dd command.

Example 1-2. Restore .img file back to the onboard flash

```
#!/bin/sh
echo timer > /sys/class/leds/beaglebone\:green\:usr0/trigger
dd if=/mnt/BeagleBoneBlack-eMMC-image-XXXXX.img of=/dev/mmcblk1 bs=10M
sync
echo default-on > /sys/class/leds/beaglebone\:green\:usr0/trigger
```

To copy the saved image back, eject the microSD card from your computer, insert it into the powered-off Bone, and apply power. Remember, because you are programming the onboard flash, you will need to use an external 5 V power supply. The Bone will boot up from the microSD card and write the backup files to the onboard flash.

1.13 Updating the Onboard Flash

Problem

You want to update the onboard flash rather than boot from the microSD card.

Solution

 At the time of this writing, these instructions were up-to-date. Go to *http://beagleboard.org/latest-images* for the latest instructions.

If you want to use the onboard flash, you need to repeat the steps in Recipe 1.10, substituting BBB-eMMC-flasher-debian-7.7-lxde-4gb-armhf-2014-11-11-4gb.img.xz for lxde-4gb/bone-debian-7.7-lxde-4gb-armhf-2014-11-11-4gb.img.xz.

That is, download, uncompress, and copy to a microSD card by using the following commands:

```
host$ wget http://rcn-ee.net/deb/testing/2014-11-11/\
BBB-eMMC-flasher-debian-7.7-lxde-4gb-armhf-2014-11-11-4gb.img.xz
host$ unxz BBB-eMMC-flasher-debian-7.7-lxde-4gb-armhf-2014-11-11-4gb.img.xz
host$ sudo dd if=BBB-eMMC-flasher-debian-7.7-lxde-4gb-armhf-2014-11-11-4gb.img \
    of=/dev/sdb bs=8M
```

Again, you'll put the microSD card in the Bone and boot. However, there is one important difference: *you must be powered from an external 5 V source.* The flashing process requires more current than what typically can be pulled from USB.

If you write the onboard flash, *be sure to power the Bone from an external 5 V source.* The USB might not supply enough current.

When you boot from the microSD card, it will copy the image to the onboard flash. When all four USER LEDs turn off (in some versions, they all turn on), you can power down the Bone and remove the microSD card. The next time you power up, the Bone will boot from the onboard flash.

CHAPTER 2

Sensors

2.0 Introduction

In this chapter, you will learn how to sense the physical world with BeagleBone Black.
Various types of electronic sensors, such as cameras and microphones, can be connected to the Bone using one or more interfaces provided by the standard USB 2.0
host port, as shown in Figure 2-1.

Figure 2-1. The USB 2.0 host port

The two 46-pin cape headers (called P8 and P9) along the long edges of the board (Figure 2-2) provide connections for cape add-on boards, digital and analog sensors, and more.

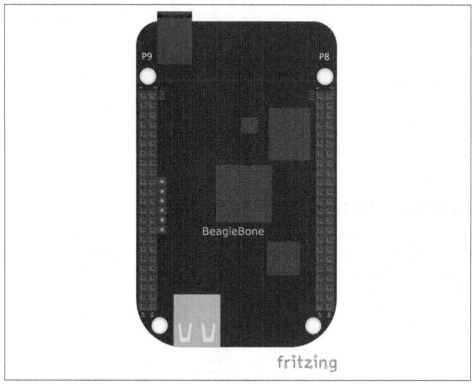

Figure 2-2. The P8 and P9 cape headers

The simplest kind of sensor provides a single digital status, such as off or on, and can be handled by an *input mode* of one of the Bone's 65 general-purpose input/output (GPIO) pins. More complex sensors can be connected by using one of the Bone's seven analog-to-digital converter (ADC) inputs or several I²C buses.

Chapter 3 discusses some of the *output mode* usages of the GPIO pins.

All these examples assume that you know how to edit a file (Recipe 1.6) and run it, either within the Cloud9 integrated development environment (IDE) or from the command line (Recipe 5.3).

2.1 Choosing a Method to Connect Your Sensor

Problem

You want to acquire and attach a sensor and need to understand your basic options.

Solution

Figure 2-3 shows many of the possibilities for connecting a sensor.

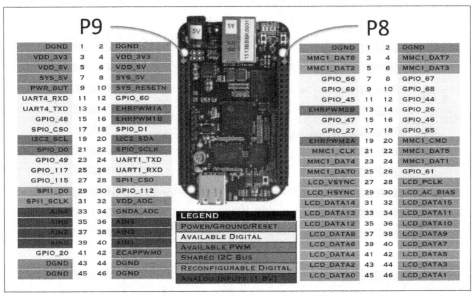

Figure 2-3. Some of the many sensor connection options on the Bone

Choosing the simplest solution available enables you to move on quickly to addressing other system aspects. By exploring each connection type, you can make more informed decisions as you seek to optimize and troubleshoot your design.

Discussion

Trying to determine which connection option to use can sometimes be tricky and often requires you to read the sensor datasheet. The names of the individual signals can give you some clues, but you should take care to note the sensor's voltage and current requirements, as well as input or output mode, to avoid damaging either the Bone or the sensor. An active sensor will require power and ground connections, whereas a passive sensor will alter an electrical property of its own, such as resistance, based on what it senses.

The digital I/O pins are very flexible. Each can be configured to have an internal pull-up or pull-down resistor (I'll explain what these are shortly) and support up to eight different major modes of operation. Each supports a GPIO mode, in which they can simply output a 0 or 3.3 V level or input a 0 or 3.3 V level as a 0 or 1. Any digital I/O pin in GPIO mode can generate an interrupt upon transitioning from one level to the other. Some of the other major modes, including eight of the digital I/O pins, can be configured to generate a pulse width modulated (PWM) signal with a frequency below once per second or above 25 MHz. There are UARTS for serial I/O and a couple of I²C and SPI ports.

The seven analog input pins do not change modes and always read an analog voltage between 0 and 1.8 V. They also include a reference ground and 1.8 V for use with your circuits.

 Keep in mind that some off-the-shelf capes (Chapter 9) might use different pins, such as the BB View LCD cape (*http://bit.ly/ 1B4r7bY*), which uses some of these analog inputs for the touch screen and many I/Os to interface to the display.

For many sensors, the simplest solution might be to connect an off-the-shelf commercial USB sensor, such as a USB microphone. The Bone supports many USB devices, as long as a Linux driver is available. If you are cost sensitive and not into writing Linux device drivers, it is best to search the Web to find out if your USB device is supported before you buy it.

Recipes in this chapter not using a simple USB-connected peripheral are built around a breadboard (Recipe 1.5). The sensor is plugged in to the breadboard or used with a breakout board, some jumper wires, and possibly some resistors. The software for the solution is written in JavaScript using the BoneScript (*http://bit.ly/1b21TGh*) library and real-time solutions discussed in Chapter 8. This approach enables you to become familiar with how the sensor is wired and data is gathered quickly before moving on to a more permanent solution, such as a printed circuit board (PCB) and Linux device driver, if either of those is ever needed.

2.2 Input and Run a JavaScript Application for Talking to Sensors

Problem

You have your sensors all wired up and your Bone booted up, and you need to know how to enter and run your code.

Solution

You are just a few simple steps from running any of the recipes in this book.

1. Plug your Bone into a host computer via the USB cable (Recipe 1.2).

2. Start Cloud9 (Recipe 1.6).

3. In the bash tab (as shown in Figure 2-4), run the following commands:

```
root@beaglebone:/var/lib/cloud9# cd
root@beaglebone:~#
```

Figure 2-4. Entering commands in the Cloud9 bash tab

Here, we issued the *change directory* (cd) command without specifying a target directory. By default, it takes you to your home directory. Notice that the prompt has changed to reflect the change. The path changed from /var/lib/cloud9 to ~. The ~ is a shorthand for your home directory.

If you log in as root, your home directory is */root*. That is, anyone can cd /root to get into your home directory (though directories are initially locked). If you log in as debian, your home is */home/ debian*. If you were to create a new user called newuser, that user's home would be */home/newuser*. By default, all non-root (non-superuser) users have their home directories in */home*.

The following commands create a new directory for the sensor recipes, change to it, and use touch to create an empty file called *pushbutton.js*:

```
root@beaglebone:/var/lib/cloud9# cd /root
root@beaglebone:~# mkdir boneSensors
root@beaglebone:~# cd boneSensors
root@beaglebone:~/boneSensors# touch pushbutton.js
```

Now, add recipe code to the newly created *pushbutton.js* file to enable it to run:

1. In Cloud9, in the Workspace browser on the left, go to the Preferences wheel and select Add Home to Favorites (Figure 1-22). Now, your home directory will show up in the Workspace file tree.

2. In the Workspace browser, expand root.

3. You will see your newly created *boneSensors* directory; expand it. You will see your empty *pushbutton.js*

4. Double-click the *pushbutton.js* file to open it.

5. Paste the code for the recipe you want to run. This example uses Example 2-1.

6. Press ^S (Ctrl-S) to save the file. (You can also go to the File menu in Cloud9 and select Save to save the file, but Ctrl-S is easier.)

7. In the bash tab, enter the following commands:

```
root@beaglebone:~/boneSensors# chmod +x pushbutton.js
root@beaglebone:~/boneSensors# ./pushbutton.js
Interrupt handler attached
x.value = 1
x.err   = undefined
x.value = 0
x.err   = undefined
x.value = 1
x.err   = undefined
^C
```

This process will work for any script in this book.

In this example, I've called b.pinMode() asynchronously and performed the b.attachInterrupt() call within the callback function. There are cases where BoneScript version 0.2.4 might not completely finish setting up a pin's mode when called synchronously. That might result in needing to run a program twice to overcome the error. Using asynchronous calls is the preferred strategy to avoid this issue.

Discussion

The chmod command changes the mode of the file *pushbutton.js* file to allow it to execute. The second command runs it.

The file you created is stored on the Bone. It's a good idea to back up your files. If your host computer is running Linux, the following commands will copy the contents of *boneSensors* to your host (be sure to subsitute your login name for yoder):

```
root@beaglebone:~/boneSensors# cd ..
root@beaglebone:~# scp -r boneSensors/ yoder@192.168.7.1:.
The authenticity of host '192.168.7.1 (192.168.7.1)' can't be established.
ECDSA key fingerprint is 54:ce:02:e5:83:3f:01:b3:bc:fd:43:09:08:d4:97:xx.
Are you sure you want to continue connecting (yes/no)? yes
Warning: Permanently added '192.168.7.1' (ECDSA) to the list of known hosts.
yoder@192.168.7.1's password:
pushbutton.js              100%  282     0.3KB/s   00:00
```

This uses the secure copy (scp) command to copy boneSensors to your home directory on your host computer.

 If scp doesn't work, you'll have to install a Secure Shell (SSH) server on your host:

```
host$ sudo apt-get install openssh-server
```

Now you can run scp from your Bone.

Or, if you don't want to install the SSH server on your host, you can back up the Bone's files from your host:

```
host$ scp -r root@192.168.7.2:boneSensors .
Warning: Permanently added '192.168.7.2' (ECDSA)
to the list of known hosts.
pushbutton.js              100%  282     0.3KB/s   00:00
```

2.3 Reading the Status of a Pushbutton or Magnetic Switch (Passive On/Off Sensor)

Problem

You want to read a pushbutton, a magnetic switch, or other sensor that is electrically open or closed.

Solution

Connect the switch to a GPIO pin and use BoneScript `pinMode()` and `attachInterrupt()` functions.

To make this recipe, you will need:

- Breadboard and jumper wires (see "Prototyping Equipment" on page 316)
- Pushbutton switch (see "Miscellaneous" on page 318)
- Magnetic reed switch (optional, see "Miscellaneous" on page 318)

You can wire up either a pushbutton, a magnetic reed switch, or both on the Bone, as shown in Figure 2-5.

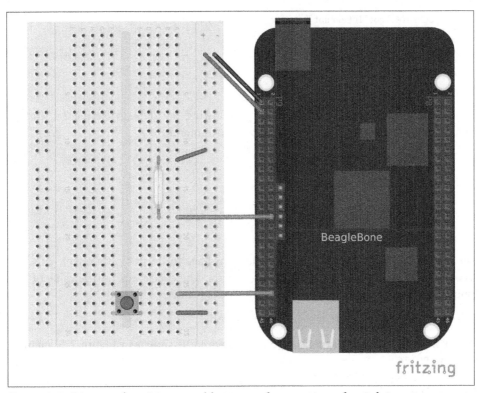

Figure 2-5. Diagram for wiring a pushbutton and magnetic reed switch input

The code in Example 2-1 reads GPIO port `P9_42`, which is attached to the pushbutton.

Example 2-1. Monitoring a pushbutton using a callback function (pushbutton.js)

```
#!/usr/bin/env node
var b = require('bonescript');
var button = 'P9_42';

b.pinMode(button, b.INPUT, 7, 'pulldown', 'fast', doAttach);

function doAttach(x) {
    if(x.err) {
        console.log('x.err = ' + x.err);
        return;
    }
    b.attachInterrupt(button, true, b.CHANGE, printStatus);
}

function printStatus(x) {
    if(x.attached) {
```

```
        console.log("Interrupt handler attached");
        return;
    }
    console.log('x.value = ' + x.value);
    console.log('x.err   = ' + x.err);
}
```

Put this code in a file called *pushbutton.js* following the steps in Recipe 2.2. In the Cloud9 bash tab, run it by using the following commands:

```
bone# chmod +x ./pushbutton.js
bone# ./pushbutton.js
Interrupt handler attached
x.value = 1
x.err   = undefined
x.value = 0
x.err   = undefined
```

The chmod command makes it executable (you have to do this only once), and the second command runs it. Try pushing the button. The code waits for the value of the input to change; when it changes, the new value is printed out.

You will have to press ^C (Ctrl-C) to stop the code.

If you want to use the magnetic reed switch wired as shown in Figure 2-5, change P9_42 to P9_26.

Discussion

In BoneScript, the attachInterrupt() command is used to trigger a function upon status change of a GPIO port. The last argument (printStatus) is the name of the function to call when the value changes. If you had other work for your Bone to do, you could list those commands after the attachInterrupt() command, and they would execute while BoneScript is waiting for activity on the GPIO port. In this simple example, there is nothing else to do, but the program won't exit, due to the attached event handler. That's why you have to press ^C (Ctrl-C).

Example 2-2 shows a more traditional way of reading a GPIO port. Here, the digital Read() command returns the value of the port synchronously. The code reads the switch only once, so to see a difference in the output, run it once with the button pushed and run it a second time without pressing the button.

Example 2-2. Reading a pushbutton by returning a value (pushbutton2.js)

```
#!/usr/bin/env node
var b = require('bonescript');
var button = 'P9_42';
var state;       // State of pushbutton
```

```
b.pinMode(button, b.INPUT, 7, 'pulldown');

state = b.digitalRead(button);
console.log('button state = ' + state);
```

This traditional, synchronous style is useful for small programs, in which you don't need to be concerned about many different possible events. Asynchronous code helps free up the processor and operating system to handle other events as they occur. It is also possible to use digitalRead() asynchronously by providing a callback function, such as in Example 2-3.

Example 2-3. Reading a pushbutton once using a callback function
(pushbutton_digitalRead.js)

```
#!/usr/bin/env node
var b = require('bonescript');
var button = 'P9_42';

b.pinMode(button, b.INPUT, 7, 'pulldown');
b.digitalRead(button, printStatus);

function printStatus(x) {
    console.log('x.value = ' + x.value);
    console.log('x.err   = ' + x.err);
}
```

What do you get if you read a GPIO port that has nothing attached to it: 1 or 0? That depends on many things. It's best not to count on either 1 or 0. A standard approach is to tie the port to either ground or 3.3 V through a 1 kΩ (or so) resistor. If nothing else is attached, the resistor will pull the input to either 0 V (ground) or 3.3 V. This is called a *pull-up* or *pull-down* resistor, depending on the power supply to which it's attached.

You can then use a switch to connect the port to the opposite value. In the previous section, the Bone was configured to use an internal pull-down resistor. When the button isn't pushed, the GPIO read 0 because the pull-down resistor held it at 0 V. When the switch is pushed, the GPIO port is attached to the 3.3 V source and a value of 1 is read.

Note that our examples set the fourth argument of pinMode() to be pulldown, to use a resistor internal to the Bone to pull the pin down to ground when the switch is open. It is also possible to configure for a pullup resistor that will pull the pin up to 3.3 V when the switch is open. Notice that the top pushbutton in Figure 2-6 is wired to ground and the bottom is wired to 3.3 V.

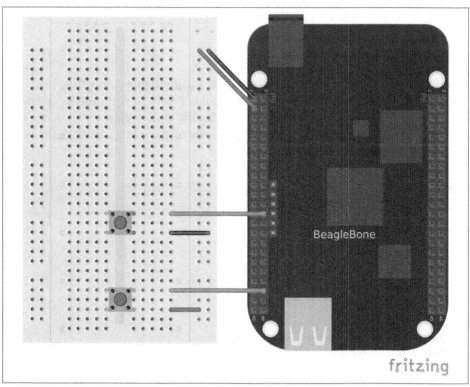

Figure 2-6. Pushbutton with Bone configured for pull-up resistor

The code in Example 2-4 configures the P9_26 GPIO port to have a pull-up resistor. Therefore, if the top button isn't pushed, a value of 1 will be read. When the top button is pushed, the GPIO port is attached to ground and a 0 is read. GPIO port P9_42 works as in Example 2-3.

Example 2-4. Reading a pushbutton with a pull-up and pull-down resistor (pushbuttonPullup.js)

```
#!/usr/bin/env node
var b = require('bonescript');
var buttonTop = 'P9_26';

b.pinMode(buttonTop, b.INPUT, 7, 'pullup', 'fast', doAttachTop);

function doAttachTop(x) {
    if(x.err) {
        console.log('x.err = ' + x.err);
        return;
    }
    b.attachInterrupt(buttonTop, true, b.CHANGE, printStatus);
```

```
}

var buttonBot = 'P9_42';
b.pinMode(buttonBot, b.INPUT, 7, 'pulldown', 'fast', doAttachBot);

function doAttachBot(x) {
    if(x.err) {
        console.log('x.err = ' + x.err);
        return;
    }
    b.attachInterrupt(buttonBot, true, b.CHANGE, printStatus);
}

function printStatus(x) {
    if(x.attached) {
        console.log("Interrupt handler attached");
        return;
    }

    console.log('x.value = ' + x.value);
    console.log('x.err   = ' + x.err);
}
```

2.4 Reading a Position, Light, or Force Sensor (Variable Resistance Sensor)

Problem

You have a variable resistor, force-sensitive resistor, flex sensor, or any of a number of other sensors that output their value as a variable resistance, and you want to read their value with the Bone.

Solution

Use the Bone's analog-to-digital converters (ADCs) and a resistor divider circuit to detect the resistance in the sensor.

The Bone has seven built-in analog inputs that can easily read a resistive value. Figure 2-7 shows them on the lower part of the P9 header.

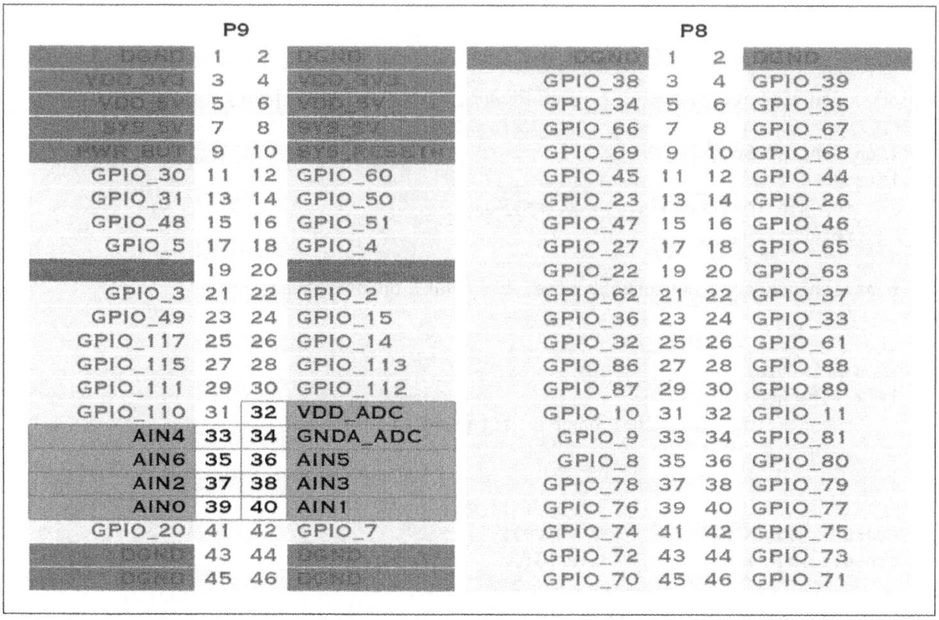

P9				P8			
DGND	1	2	DGND	DGND	1	2	DGND
VDD_3V3	3	4	VDD_3V3	GPIO_38	3	4	GPIO_39
VDD_5V	5	6	VDD_5V	GPIO_34	5	6	GPIO_35
SYS_5V	7	8	SYS_5V	GPIO_66	7	8	GPIO_67
PWR_BUT	9	10	SYS_RESETN	GPIO_69	9	10	GPIO_68
GPIO_30	11	12	GPIO_60	GPIO_45	11	12	GPIO_44
GPIO_31	13	14	GPIO_50	GPIO_23	13	14	GPIO_26
GPIO_48	15	16	GPIO_51	GPIO_47	15	16	GPIO_46
GPIO_5	17	18	GPIO_4	GPIO_27	17	18	GPIO_65
	19	20		GPIO_22	19	20	GPIO_63
GPIO_3	21	22	GPIO_2	GPIO_62	21	22	GPIO_37
GPIO_49	23	24	GPIO_15	GPIO_36	23	24	GPIO_33
GPIO_117	25	26	GPIO_14	GPIO_32	25	26	GPIO_61
GPIO_115	27	28	GPIO_113	GPIO_86	27	28	GPIO_88
GPIO_111	29	30	GPIO_112	GPIO_87	29	30	GPIO_89
GPIO_110	31	32	VDD_ADC	GPIO_10	31	32	GPIO_11
AIN4	33	34	GNDA_ADC	GPIO_9	33	34	GPIO_81
AIN6	35	36	AIN5	GPIO_8	35	36	GPIO_80
AIN2	37	38	AIN3	GPIO_78	37	38	GPIO_79
AIN0	39	40	AIN1	GPIO_76	39	40	GPIO_77
GPIO_20	41	42	GPIO_7	GPIO_74	41	42	GPIO_75
DGND	43	44	DGND	GPIO_72	43	44	GPIO_73
DGND	45	46	DGND	GPIO_70	45	46	GPIO_71

Figure 2-7. Seven analog inputs on the P9 header

To make this recipe, you will need:

- Breadboard and jumper wires (see "Prototyping Equipment" on page 316)
- 10 kΩ trimpot (see "Resistors" on page 316) or
- Flex resistor (optional, see "Resistors" on page 316)
- 22 kΩ resistor (see "Resistors" on page 316)

A variable resistor with three terminals

Figure 2-8 shows a simple variable resistor (trimpot) wired to the Bone. One end terminal is wired to the ADC 1.8 V power supply on pin P9_32, and the other end terminal is attached to the ADC ground (P9_34). The middle terminal is wired to one of the seven analog-in ports (P9_36).

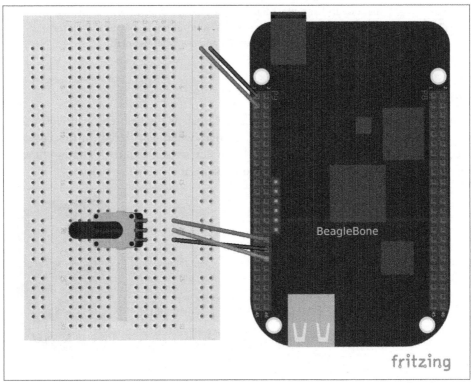

Figure 2-8. Wiring a 10kΩ variable resistor (trimpot) to an ADC port

Example 2-5 shows the BoneScript code used to read the variable resistor. Add the code to a file called *analogIn.js* and run it; then change the resistor and run it again. The voltage read will change.

Example 2-5. Reading an analog voltage (analogIn.js)

```
#!/usr/bin/env node
var b = require('bonescript');

b.analogRead('P9_36', printStatus);

function printStatus(x) {
    console.log('x.value = ' + x.value.toFixed(3));
    console.log('x.err   = ' + x.err);
}
```

This code in Example 2-5 uses `.toFixed(3)` to print only three digits past the decimal. Otherwise, the output could have many extra digits. Try removing `.toFixed(3)` and see what happens.

A variable resistor with two terminals

Some resistive sensors have only two terminals, such as the flex sensor in Figure 2-9. The resistance between its two terminals changes when it is flexed. In this case, we need to add a fixed resistor in series with the flex sensor. Figure 2-9 shows how to wire in a 22 kΩ resistor to give a voltage to measure across the flex sensor.

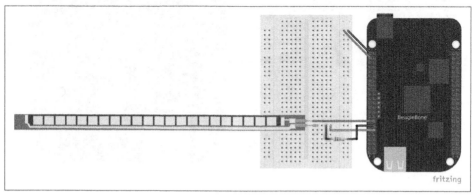

Figure 2-9. Reading a two-terminal flex resistor

The code in Example 2-5 also works for this setup.

Discussion

The *analog-in* pins on the Bone read a voltage between 0 and 1.8 V. (This is different from the *digital-in* pins, which read either 0 or 3.3 V.) A variable resistor (as shown in Figure 2-10) has a fixed resistor between two terminals and a wiper that can be moved between them. As the wiper moves, the resistance between it and the ends changes—that is, the resistance drops between the wiper and the end to which it is getting closer.

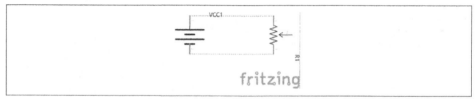

Figure 2-10. Schematic for a variable resistor

This example puts the analog 1.8 V reference voltage (which appears on P9_32) and the analog ground (which appears on P9_34) across the fixed resistor. As the resistor is turned, the voltage read on the wiper varies from 0 to 1.8 V, which is read by the *analog-in* pin on P9_36.

The flex sensor has no fixed resistor; rather, it has only two terminals whose resistance across them will vary when it is flexed. In Figure 2-9, a 22 kΩ resistor is used to form the circuit needed to get a variable voltage to read.

2.5 Reading a Distance Sensor (Analog or Variable Voltage Sensor)

Problem

You want to measure distance with a LV-MaxSonar-EZ1 Sonar Range Finder (*http:// bit.ly/1Mt5Elr*), which outputs a voltage in proportion to the distance.

Solution

To make this recipe, you will need:

- Breadboard and jumper wires (see "Prototyping Equipment" on page 316)
- LV-MaxSonar-EZ1 Sonar Range Finder (see "Miscellaneous" on page 318)

All you have to do is wire the EZ1 to one of the Bone's *analog-in* pins, as shown in Figure 2-11. The device outputs ~6.4 mV/in when powered from 3.3 V.

Make sure not to apply more than 1.8 V to the Bone's *analog-in* pins, or you will likely damage them. In practice, this circuit should follow that rule.

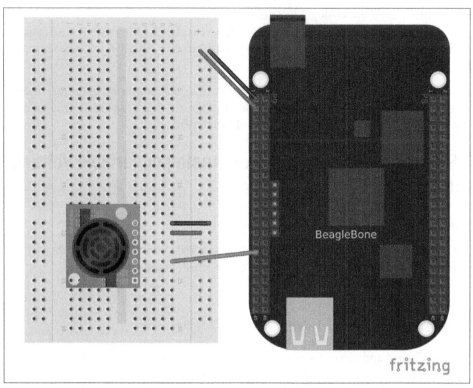

Figure 2-11. Wiring the LV-MaxSonar-EZ1 Sonar Range Finder to the P9_33 analog-in port

Example 2-6 shows the code that reads the sensor at a fixed interval.

Example 2-6. Reading an analog voltage (ultrasonicRange.js)

```
#!/usr/bin/env node
var b = require('bonescript');
var ms = 250;   // Time in milliseconds

setInterval(readRange, ms);

function readRange() {
    b.analogRead('P9_33', printStatus);
}
function printStatus(x) {
    console.log('x.value = ' + x.value);
    console.log('Distance= ' + x.value * 1.8/0.0064);
}
```

Discussion

The LV-MaxSonar-EZ1 outputs distance measured in several ways: RS232 serial, a PWM signal, and an analog voltage. The easiest one to use is the analog-in, because the Bone has several analog-in ports.

If you take the analog voltage*1.8/0.0064, you'll get the distance in inches.

The setInterval(readRange, ms); statement in Example 2-6 causes the read Range() function to be called every 250 ms. The readRange() function reads the analog port and calls printStatus() once the value is ready. printStatus() then prints the value read.

The built-in ADC on the Bone is a 12-bit, 200 k samples-per-second successive approximation converter with eight channels. Seven of the channels are available on the Bone's *analog-in* pins.

2.6 Reading a Distance Sensor (Variable Pulse Width Sensor)

Problem

You want to use a HC-SR04 Ultrasonic Range Sensor with BeagleBone Black.

Solution

The HC-SR04 Ultrasonic Range Sensor (shown in Figure 2-12) works by sending a trigger pulse to the *Trigger* input and then measuring the pulse width on the *Echo* output. The width of the pulse tells you the distance.

Figure 2-12. HC-SR04 Ultrasonic range sensor

To make this recipe, you will need:

- Breadboard and jumper wires (see "Prototyping Equipment" on page 316)
- 10 kΩ and 20 kΩ resistors (see "Resistors" on page 316)
- HC-SR04 Ultrsonic Range Sensor (see "Miscellaneous" on page 318)

Wire the sensor as shown in Figure 2-13. Note that the HC-SR04 is a 5 V device, so the *banded* wire (running from P9_7 on the Bone to VCC on the range finder) attaches the HC-SR04 to the Bone's 5 V power supply.

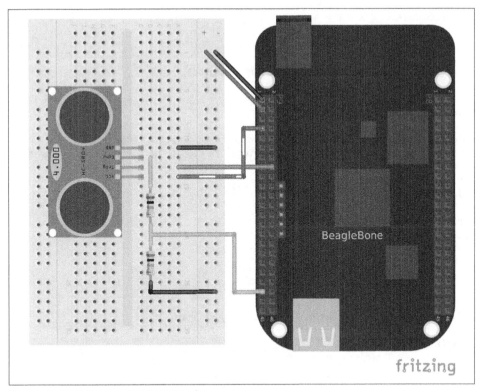

Figure 2-13. Wiring an HC-SR04 ultrasonic sensor

Example 2-7 shows BoneScript code used to drive the HC-SR04.

Example 2-7. Driving a HC-SR04 ultrasound sensor (hc-sr04-ultraSonic.js)

```
#!/usr/bin/env node

// This is an example of reading HC-SR04 Ultrasonic Range Finder
// This version measures from the fall of the Trigger pulse
//    to the end of the Echo pulse

var b = require('bonescript');

var trigger = 'P9_16',  // Pin to trigger the ultrasonic pulse
    echo    = 'P9_41',  // Pin to measure to pulse width related to the distance
    ms = 250;           // Trigger period in ms

var startTime, pulseTime;

b.pinMode(echo,   b.INPUT, 7, 'pulldown', 'fast', doAttach);
function doAttach(x) {
    if(x.err) {
```

```
        console.log('x.err = ' + x.err);
        return;
    }
    // Call pingEnd when the pulse ends
    b.attachInterrupt(echo, true, b.FALLING, pingEnd);
}

b.pinMode(trigger, b.OUTPUT);

b.digitalWrite(trigger, 1);    // Unit triggers on a falling edge.
                               // Set trigger to high so we call pull it low later

// Pull the trigger low at a regular interval.
setInterval(ping, ms);

// Pull trigger low and start timing.
function ping() {
    // console.log('ping');
    b.digitalWrite(trigger, 0);
    startTime = process.hrtime();
}

// Compute the total time and get ready to trigger again.
function pingEnd(x) {
    if(x.attached) {
        console.log("Interrupt handler attached");
        return;
    }
    if(startTime) {
        pulseTime = process.hrtime(startTime);
        b.digitalWrite(trigger, 1);
        console.log('pulseTime = ' + (pulseTime[1]/1000000-0.8).toFixed(3));
    }
}
```

This code is more complex than others in this chapter, because we have to tell the device when to start measuring and time the return pulse.

Discussion

The HC-SR04 works by sending a 40 kHz ultrasonic pulse out when a 10 µs (or longer) pulse appears on its *Trigger* input. The *Echo* output produces a pulse whose length is in proportion to the distance that was measured.

The HC-SR04 is a 5 V device that outputs a 5 V pulse on its *Echo* port, which is too much for the Bone's GPIOs, which expect 0 and 3.3 V. A simple resistor divider network is used to scale the 5 V down to 3.3 V. We use 20 kΩ and 10 kΩ resistors for the divider, but other pairs will work as long as the ratio is 2 to 1. But be careful: if you make them too big (say, 100 kΩ and 200 kΩ), the GPIO port will draw too much current through it and make the output voltage sag too much. If you make the resistors too small, you will pull a lot of current through them, wasting power.

The code in Example 2-7 works by setting `trigger` to 1 and then using `setInterval()` and `ping()` to pull `trigger` low at a regular interval and start a timer. `attachInterrupt()` and `pingEnd()` are used to wait for `echo` to fall to 0. Once it does, `pingEnd()` is called, and it computes the time since `ping()` started and sets `trigger` back to 1, so everything can start over again.

Example 2-7 uses a high-resolution timer (`hrtime()`), which returns the time in nanoseconds. (JavaScript's built-in timer gives only *ms* results.) The code converts the nanoseconds to milliseconds and then subtracts 0.8 ms. The 0.8 ms is the approximate time from when the trigger falls to when the pulse is sent out. (This is a measured result, not from the datasheet.)

But how accurate is it? I attached an oscilloscope to the *Echo* pin and measured pulses with widths from 0.5 to 5 ms. The oscilloscope measured a 25 µs standard deviation, which represents the device's repeatability and the stability of the person holding the target. With the 0.8 ms correction, the Bone appeared to measure the pulse width to within 0.2 ms or so, which is around 4 cm. Not bad for using JavaScript and probably close enough for most robotics applications.

2.7 Accurately Reading the Position of a Motor or Dial

Problem

You have a motor or dial and want to detect rotation using a rotary encoder.

Solution

Use a rotary encoder (also called a *quadrature encoder*) connected to one of the Bone's eQEP ports, as shown in Figure 2-14.

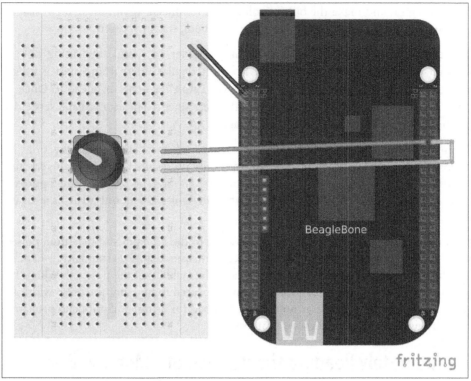

Figure 2-14. Wiring a rotary encoder using eQEP2

To make this recipe, you will need:

- Breadboard and jumper wires (see "Prototyping Equipment" on page 316)
- Rotary encoder (see "Miscellaneous" on page 318)

We are using a quadrature rotary encoder, which has two switches inside that open and close in such a manner that you can tell which way the shaft is turning. In this particular encoder, the two switches have a common lead, which is wired to ground. It also has a pushbutton switch wired to the other side of the device, which we aren't using.

Wire the encoder to P8_11 and P8_12, as shown in Figure 2-14.

BeagleBone Black has built-in hardware for reading up to three encoders. Here, we'll use the *eQEP2* encoder. To enable it, put the code from Example 2-8 in a file called *bone_eqep2b.dts*. You can do this using Cloud9 to edit files (as shown in Recipe 1.6) or use a more traditional editor (as shown in Recipe 5.9).

Example 2-8. Configuring a rotary encoder (bone_eqep2b.dts) [1]

```
/*
 * Copyright (C) 2013 Nathaniel R. Lewis - http://nathanielrlewis.com/
 *
 * This program is free software; you can redistribute it and/or modify
 * it under the terms of the GNU General Public License version 2 as
 * published by the Free Software Foundation.
 *
 * Enable eQEP2 on the Beaglebone White and Black
 * These pins don't conflict with the HDMI
 */
/dts-v1/;
/plugin/;

/ {
    compatible = "ti,beaglebone", "ti,beaglebone-black";

    /* identification */
    part-number = "bone_eqep2";
    version     = "00A0";

    fragment@0 {
        target = <&am33xx_pinmux>;
        __overlay__ {
            pinctrl_eqep2: pinctrl_eqep2_pins {
                pinctrl-single,pins = <
                        0x038 0x24  /* P8_16 = GPIO2_12 = EQEP2_index, MODE4 */
                        0x03C 0x24  /* P8_15 = GPIO2_13 = EQEP2_strobe, MODE4 */
                        0x030 0x34  /* P8_12 = GPIO2_10 = EQEP2A_in, MODE4 */
                        0x034 0x34  /* P8_11 = GPIO2_11 = EQEP2B_in, MODE4 */
                >;
            };
        };
    };

    fragment@1 {
        target = <&epwmss2>;
        __overlay__ {
            status = "okay";
        };
    };

    fragment@2 {
     target = <&eqep2>;
     __overlay__ {
        pinctrl-names = "default";
        pinctrl-0 = <&pinctrl_eqep2>;
```

---

1 *This solution first appeared on the Beagle Google Group (https://groups.google.com/forum/#!searchin/beagle board/eQep/beagleboard/Orp3tFcNgCc/mYacP_GkCQQJ).*

```
            count_mode = < 0 >;
            /* 0 - Quadrature mode, normal 90 phase offset cha & chb.
               1 - Direction mode.  cha input = clock, chb input = direction  */
            swap_inputs = < 0 >; /* Are chan A & chan B swapped? (0-no,1-yes) */
              invert_qa = < 1 >;   /* Should we invert the channel A input?  */
              invert_qb = < 1 >;   /* Should we invert the channel B input?
                  These are inverted because my encoder outputs drive transistors
                  that pull down the pins */
              invert_qi = < 0 >;   /* Should we invert the index input? */
              invert_qs = < 0 >;   /* Should we invert the strobe input? */

            status = "okay";
        };
        };
};
```

Then run the following commands:

```
bone# dtc -O dtb -o bone_eqep2b-00A0.dtbo -b 0 -@ bone_eqep2b.dts
bone# cp bone_eqep2b-00A0.dtbo /lib/firmware
bone# echo bone_eqep2b > /sys/devices/bone_capemgr.*/slots
```

This will enable *eQEP2* on pins P8_11 and P8_12.

Finally, add the code in Example 2-9 to a file named *rotaryEncoder.js* and run it.

Example 2-9. Reading a rotary encoder (rotaryEncoder.js)

```
#!/usr/bin/env node
// This uses the eQEP hardware to read a rotary encoder
// echo bone_eqep2b > $SLOTS

var b = require('bonescript'),
    fs = require('fs');

var eQEP0 = "/sys/devices/ocp.3/48300000.epwmss/48300180.eqep/",
    eQEP1 = "/sys/devices/ocp.3/48302000.epwmss/48302180.eqep/",
    eQEP2 = "/sys/devices/ocp.3/48304000.epwmss/48304180.eqep/",
    eQEP = eQEP2;

var oldData,                    // pervious data read
    period = 100;        // in ms

// Set the eEQP period, convert to ns.
fs.writeFile(eQEP+'period', period*1000000, function(err) {
        if (err) throw err;
        console.log('Period updated to ' + period*1000000);
})

// Enable
fs.writeFile(eQEP+'enabled', 1, function(err) {
        if (err) throw err;
```

```
        console.log('Enabled');
})

setInterval(readEncoder, period);    // Check state every 250 ms

function readEncoder(x) {
        fs.readFile(eQEP + 'position', {encoding: 'utf8'}, printValue);
}

function printValue(err, data) {
        if (err) throw err;
        if (oldData !== data) {
                console.log('position: '+data+' speed: '+(oldData-data));
                oldData = data;
        }
}
```

Try rotating the encoder clockwise and counter-clockwise. You'll see an output like this:

```
Period updated to 100000000
Enabled
position: 0
 speed: NaN
position: 4
 speed: -4
position: 6
 speed: -2
position: 8
 speed: -2
position: 12
 speed: -4
position: 16
 speed: -4
position: 19
 speed: -3
position: 20
 speed: -1
```

The values you get for speed and position will depend on which way you are turning the device and how quickly. You will need to press ^C (Ctrl-C) to end the program.

Discussion

Figure 2-15 shows the sequence of pulses that occur when the encoder is turned in one direction (either clockwise or counter-clockwise, depending on the encoder).

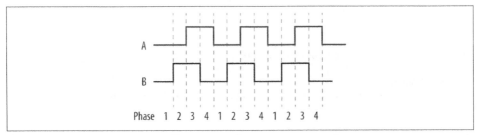

Figure 2-15. Pulses from a quadrature encoder

When rotating clockwise, the outputs are as follows:

Phase	A	B
1	0	0
2	0	1
3	1	1
4	1	0

When rotating counter-clockwise, the outputs are as follows:

Phase	A	B
4	1	0
3	1	1
2	0	1
1	0	0

The eQEP hardware watches the A and B inputs and determines which way the encoder is turning. Because this is implemented in the hardware, it will work with encoders driven by motors revolving at thousands of RPMs.

See Also

You can also measure rotation by using a variable resistor (see Figure 2-8).

2.8 Acquiring Data by Using a Smart Sensor over a Serial Connection

Problem

You want to connect a smart sensor that uses a built-in microcontroller to stream data, such as a global positioning system (GPS), to the Bone and read the data from it.

Solution

The Bone has several serial ports (UARTs) that you can use to read data from an external microcontroller included in smart sensors, such as a GPS. Just wire one up, and you'll soon be gathering useful data, such as your own location.

Here's what you'll need:

- Breadboard and jumper wires (see "Prototyping Equipment" on page 316)
- GPS receiver (see "Miscellaneous" on page 318)

Wire your GPS, as shown in Figure 2-16.

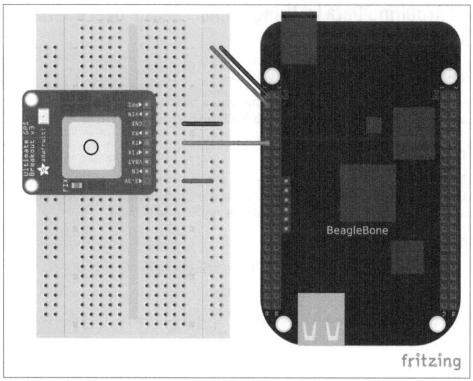

Figure 2-16. Wiring a GPS to UART 4

The GPS will produce raw National Marine Electronics Association (NMEA) data that's easy for a computer to read, but not for a human. There are many utilities to help convert such sensor data into a human-readable form. For this GPS, run the following command to load a NMEA parser:

```
bone# npm install -g nmea
```

Running the code in Example 2-10 will print the current location every time the GPS outputs it.

Example 2-10. Talking to a GPS with UART 4 (GPS.js)

```
#!/usr/bin/env node
// Install with: npm install nmea

// Need to add exports.serialParsers = m.module.parsers;
// to the end of /usr/local/lib/node_modules/bonescript/serial.js

var b = require('bonescript');
var nmea = require('nmea');
```

```
var port = '/dev/tty04';
var options = {
    baudrate: 9600,
    parser: b.serialParsers.readline("\n")
};

b.serialOpen(port, options, onSerial);

function onSerial(x) {
    if (x.err) {
        console.log('***ERROR*** ' + JSON.stringify(x));
    }
    if (x.event == 'open') {
        console.log('***OPENED***');
    }
    if (x.event == 'data') {
        console.log(String(x.data));
        console.log(nmea.parse(x.data));
    }
}
```

If you don't need the NMEA formatting, you can skip the npm part and remove the lines in the code that refer to it.

If you get an error like this

`TypeError: Cannot call method readline of undefined`

add this line to the end of file */usr/local/lib/node_modules/bone-script/serial.js*:

`exports.serialParsers = m.module.parsers;`

Discussion

The GPS outputs text every second or so. The Bone's serial port is set up to call the onSerial function whenever text is ready. When onSerial sees a data event, it converts the data to a string and prints it. The second console.log() takes the raw data and passes it through the nmea parser, which converts it to a more readable form and then prints it.

You cannot use every pin as a UART. Point your browser to the Headers page served by your Bone (*http://192.168.7.2/Support/bone101/#headers*) (or Online headers-serial page (*http://bit.ly/1B4rvqS*)) and scroll down to the UART page to see which pins work. The table is reproduced in Figure 2-17.

P9					P8			
DGND	1	2	DGND		DGND	1	2	DGND
VDD_3V3	3	4	VDD_3V3		GPIO_38	3	4	GPIO_39
VDD_5V	5	6	VDD_5V		GPIO_34	5	6	GPIO_35
SYS_5V	7	8	SYS_5V		GPIO_66	7	8	GPIO_67
PWR_BUT	9	10	SYS_RESETn		GPIO_69	9	10	GPIO_68
UART4_RXD	11	12	GPIO_60		GPIO_45	11	12	GPIO_44
UART4_TXD	13	14	GPIO_50		GPIO_23	13	14	GPIO_26
GPIO_48	15	16	GPIO_51		GPIO_47	15	16	GPIO_46
GPIO_5	17	18	GPIO_4		GPIO_27	17	18	GPIO_65
UART1_RTSn	19	20	UART1_CTSn		GPIO_22	19	20	GPIO_63
UART2_TXD	21	22	UART2_RXD		GPIO_62	21	22	GPIO_37
GPIO_49	23	24	UART1_TXD		GPIO_36	23	24	GPIO_33
GPIO_117	25	26	UART1_RXD		GPIO_32	25	26	GPIO_61
GPIO_115	27	28	GPIO_113		GPIO_86	27	28	GPIO_88
GPIO_111	29	30	GPIO_112		GPIO_87	29	30	GPIO_89
GPIO_110	31	32	VDD_ADC		UART5_CTSn+	31	32	UART5_RTSn
AIN4	33	34	GNDA_ADC		UART4_RTSn	33	34	UART3_RTSn
AIN6	35	36	AIN5		UART4_CTSn	35	36	UART3_CTSn
AIN2	37	38	AIN3		UARR5_TXD+	37	38	UART5_RXD+
AIN0	39	40	AIN1		GPIO_76	39	40	GPIO_77
GPIO_20	41	42	UART3_TXD		GPIO_74	41	42	GPIO_75
DGND	43	44	DGND		GPIO_72	43	44	GPIO_73
DGND	45	46	DGND		GPIO_70	45	46	GPIO_71

Figure 2-17. Table of UART outputs

2.9 Measuring a Temperature

Problem

You want to measure a temperature using a digital temperature sensor.

Solution

The TMP102 sensor is a common digital temperature sensor that uses a standard I^2C-based serial protocol.

To make this recipe, you will need:

- Breadboard and jumper wires (see "Prototyping Equipment" on page 316)
- Two 4.7 kΩ resistors (see "Resistors" on page 316)
- TMP102 temperature sensor (see "Integrated Circuits" on page 317)

Wire the TMP102, as shown in Figure 2-18.

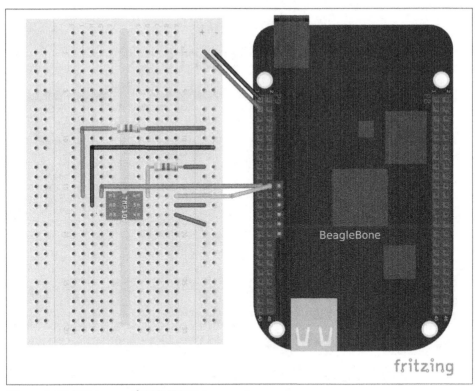

Figure 2-18. Wiring an I²C TMP102 temperature sensor

There are two I²C buses brought out to the headers. Figure 2-19 shows that you have wired your device to I²C bus 2, but watch out: the buses aren't always numbered the same. When you work with BoneScript, they are numbered 1 and 2, but if you work from the Linux command line, they are numbered 0 and 1. Confusing, huh?

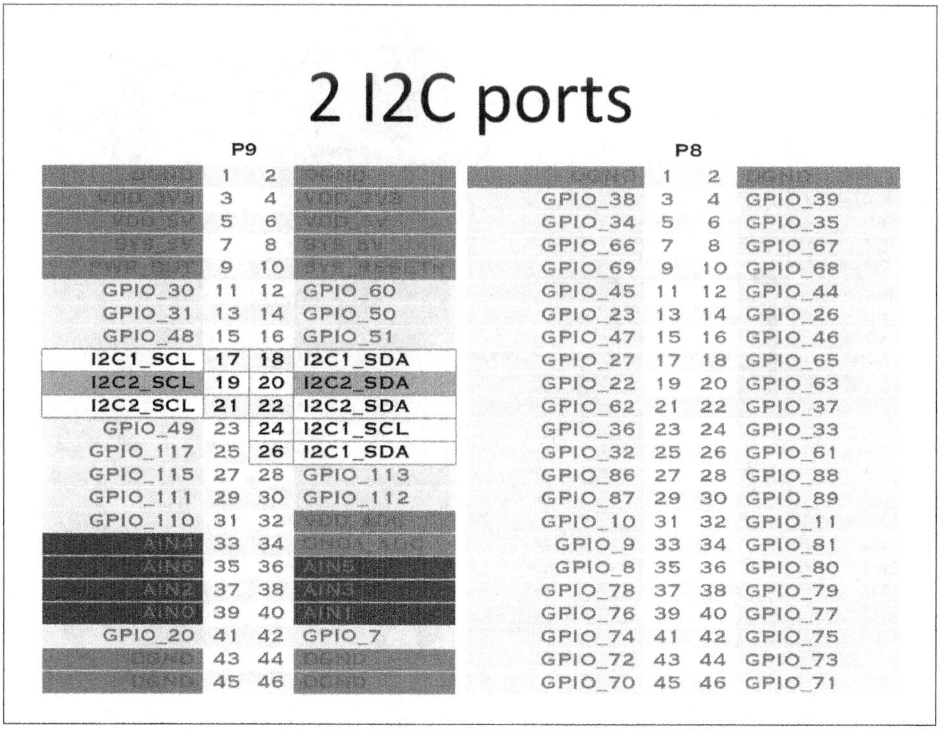

2 I2C ports

P9					P8			
DGND	1	2	DGND		DGND	1	2	DGND
VDD_3V3	3	4	VDD_3V3		GPIO_38	3	4	GPIO_39
VDD_5V	5	6	VDD_5V		GPIO_34	5	6	GPIO_35
SYS_5V	7	8	SYS_5V		GPIO_66	7	8	GPIO_67
PWR_BUT	9	10	SYS_RESETn		GPIO_69	9	10	GPIO_68
GPIO_30	11	12	GPIO_60		GPIO_45	11	12	GPIO_44
GPIO_31	13	14	GPIO_50		GPIO_23	13	14	GPIO_26
GPIO_48	15	16	GPIO_51		GPIO_47	15	16	GPIO_46
I2C1_SCL	17	18	I2C1_SDA		GPIO_27	17	18	GPIO_65
I2C2_SCL	19	20	I2C2_SDA		GPIO_22	19	20	GPIO_63
I2C2_SCL	21	22	I2C2_SDA		GPIO_62	21	22	GPIO_37
GPIO_49	23	24	I2C1_SCL		GPIO_36	23	24	GPIO_33
GPIO_117	25	26	I2C1_SDA		GPIO_32	25	26	GPIO_61
GPIO_115	27	28	GPIO_113		GPIO_86	27	28	GPIO_88
GPIO_111	29	30	GPIO_112		GPIO_87	29	30	GPIO_89
GPIO_110	31	32	VDD_ADC		GPIO_10	31	32	GPIO_11
AIN4	33	34	GNDA_ADC		GPIO_9	33	34	GPIO_81
AIN6	35	36	AIN5		GPIO_8	35	36	GPIO_80
AIN2	37	38	AIN3		GPIO_78	37	38	GPIO_79
AIN0	39	40	AIN1		GPIO_76	39	40	GPIO_77
GPIO_20	41	42	GPIO_7		GPIO_74	41	42	GPIO_75
DGND	43	44	DGND		GPIO_72	43	44	GPIO_73
DGND	45	46	DGND		GPIO_70	45	46	GPIO_71

Figure 2-19. Table of I²C outputs

Once the I²C device is wired up, you can use a couple handy I²C tools to test the device. Because these are Linux command-line tools, you have to use 1 as the bus number. i2cdetect, shown in Example 2-11, shows which I²C devices are on the bus. The -r flag indicates which bus to use. Our TMP102 is appearing at address 0x49. You can use the i2cget command to read the value. It returns the temperature in hexidecimal and degrees C. In this example, 0x18 = 24°C, which is 75.2°F. (Hmmm, the office is a bit warm today.) Try warming up the TMP102 with your finger and running i2cget again.

Example 2-11. I²C tools

```
bone# i2cdetect -y -r 1
     0  1  2  3  4  5  6  7  8  9  a  b  c  d  e  f
00:          -- -- -- -- -- -- -- -- -- -- -- -- --
10: -- -- -- -- -- -- -- -- -- -- -- -- -- -- -- --
20: -- -- -- -- -- -- -- -- -- -- -- -- -- -- -- --
30: -- -- -- -- -- -- -- -- -- -- -- -- -- -- -- --
40: -- -- -- -- -- -- -- -- -- 49 -- -- -- -- -- --
50: -- -- -- -- UU UU UU UU -- -- -- -- -- -- -- --
60: -- -- -- -- -- -- -- -- -- -- -- -- -- -- -- --
```

```
70: -- -- -- -- -- -- -- --

bone# i2cget -y 1 0x49
0x18
```

Example 2-12 shows how to read the TMP102 from BoneScript.

Example 2-12. Reading an I²C device (i2cTemp.js)

```
#!/usr/bin/env node

var b = require('bonescript');
var bus = '/dev/i2c-2'                              ❶
var TMP102 = 0x49;                                  ❷

b.i2cOpen(bus, TMP102);                             ❸
b.i2cReadByte(bus, onReadByte);                     ❹

function onReadByte(x) {                            ❺
    if (x.event == 'callback') {
        console.log('onReadByte: ' + JSON.stringify(x));  ❻
        console.log(x.res*9/5+32 + 'F');            ❼
    }
}
```

❶ This line states which bus to use. The last digit gives the BoneScript bus number.

❷ This gives the address of the device on the bus.

❸ This line opens the device. All I²C commands that follow will apply to this bus and device.

❹ This line reads a byte from the device. The default is to read from address 0 of the selected device. Address 0 for the TMP102 is the current temperature. As soon as the register is read, the onReadByte() function is called.

❺ The value, x, passed to onReadByte() is an object. The .event field of the object informs us if we have a callback.

❻ If this is a callback, use console.log to display the contents of object x.

❼ Finally, print the result field, x.res, after converting it to degrees F.

Run the code by using the following command:

```
bone# ./i2cTemp.js
onReadByte: {"err":null,"res":24,"event":"callback"}
75.2F
```

Discussion

You can easily modify this recipe to call `b.i2cReadByte(bus, onReadByte)` everytime you want to read a new temperature. You can also adapt it to read other I²C devices just by changing the bus and address (`0x49` in this example) to another I²C device. In fact, multiple I²C devices can be on the same bus at that same time and will respond only to their addresses.

2.10 Reading Temperature via a Dallas 1-Wire Device

Problem

You want to measure a temperature using a Dallas Semiconductor DS18B20 temperature sensor.

Solution

The DS18B20 is an interesting temperature sensor that uses Dallas Semiconductor's 1-wire interface. The data communication requires only one wire! (However, you still need wires from ground and 3.3 V.) You can wire it to any GPIO port.

To make this recipe, you will need:

- Breadboard and jumper wires (see "Prototyping Equipment" on page 316)
- 4.7 kΩ resistor (see "Resistors" on page 316)
- DS18B20 1-wire temperature sensor (see "Integrated Circuits" on page 317)

Wire up as shown in Figure 2-20.

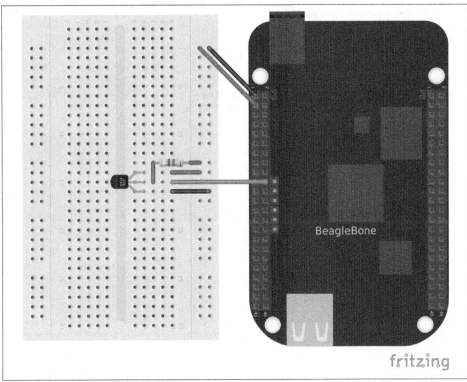

Figure 2-20. Wiring a Dallas 1-Wire temperature sensor [2]

Add the code in Example 2-13 to a file called *BB-W1-00A0.dts.*

Example 2-13. Reading a temperature with a DS18B20 (BB-W1-00A0.dts)

```
/dts-v1/;
/plugin/;

/ {
    compatible = "ti,beaglebone", "ti,beaglebone-black";

    part-number = "BB-W1";
    version = "00A0";

    /* state the resources this cape uses */
    exclusive-use =
        /* the pin header uses */
        "P9.20",
```

---

[2] *This solution, written by Elias Bakken (@AgentBrum), originally appeared on Hipstercircuits (http://bit.ly/ 1FaRbbK).*

```
        /* the hardware IP uses */
        "gpio0_12";

    fragment@0 {
        target = <&am33xx_pinmux>;
        __overlay__ {
            dallas_w1_pins: pinmux_dallas_w1_pins {
                pinctrl-single,pins = < 0x150 0x37 >;
            };
        };
    };

    fragment@1 {
        target = <&ocp>;
        __overlay__ {
            onewire@0 {
                compatible      = "w1-gpio";
                pinctrl-names   = "default";
                pinctrl-0       = <&dallas_w1_pins>;
                status          = "okay";

                gpios = <&gpio1 2 0>;
            };
        };
    };
};
```

Then run the following commands:

```
bone# dtc -O dtb -o BB-W1-00A0.dtbo -b 0 -@ BB-W1-00A0.dts
bone# cp BB-W1-00A0.dtbo /lib/firmware/
bone# echo BB-W1 > /sys/devices/bone_capemgr.*/slots
```

Now run the following command to discover the serial number on your device:

```
bone# ls /sys/bus/w1/devices/
28-00000114ef1b   28-00000128197d   w1_bus_master1
```

This shows the serial numbers for all the devices.

Finally, add the code in Example 2-14 in to a file named *onewire.js*, edit the path assigned to w1 so that the path points to your device, and then run it.

Example 2-14. Reading a temperature with a DS18B20 (onewire.js)

```
#!/usr/bin/env node
var b = require('bonescript');

var w1="/sys/bus/w1/devices/28-00000114ef1b/w1_slave"

setInterval(getTemp, 1000);     // read temperatue every 1000ms

function getTemp() {
```

```
    b.readTextFile(w1, printStatus);
}

function printStatus(x) {
    console.log('x.data = ' + x.data);
    console.log('x.err  = ' + x.err);
}
```

Discussion

Each temperature sensor has a unique serial number, so you can have several all sharing the same data line.

2.11 Sensing All Sorts of Things with SensorTag via Bluetooth v4.0

Problem

You have a TI SensorTag, and you want to interface it to BeagleBone Black via Bluetooth Low Energy (BLE).

Solution

TI's SensorTag (*http://bit.ly/1C58WIN*) (shown in Figure 2-21) combines six sensors (temperature, humidity, accelerometer, pressure, magnetometer, and gyroscope) in one package that interfaces via Bluetooth Low Energy (*http://bit.ly/1EzMo4x*).

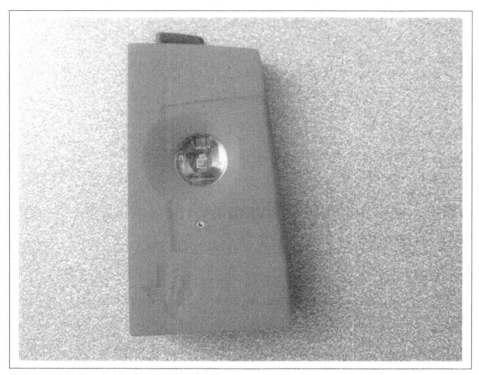

Figure 2-21. SensorTag

To make this recipe, you will need:

- BLE USB dongle (see "Miscellaneous" on page 318)
- SensorTag (see "Miscellaneous" on page 318)
- 5 V adapter for the Bone

Power up your Bone using the 5 V adapter. You need the adapter because the BLE dongle needs extra power for the radios it contains. After it is booted up, log in (Recipe 5.3) and run the following commands:

```
bone# apt-get install libbluetooth-dev
bone# npm install -g sensortag
```

This installs the Bluetooth tools and the JavaScript library to talk to it.

Add the code in Example 2-15 to a file called *sensorTag.js* and run it.

Example 2-15. Code for reading the temperature from a SensorTag (sensorTag.js)

```
#!/usr/bin/env node
// From: https://github.com/sandeepmistry/node-sensortag
```

```
// Reads temperature

var util = require('util');                    ❶
var async = require('async');
var SensorTag = require('sensortag');
var fs = require('fs');

console.log("Be sure sensorTag is on");

SensorTag.discover(function(sensorTag) {       ❷
  console.log('sensorTag = ' + sensorTag);
  sensorTag.on('disconnect', function() {      ❸
    console.log('disconnected!');
    process.exit(0);
  });

  async.series([                               ❹
      function(callback) {
        console.log('connect');                ❺
        sensorTag.connect(callback);
      },
      function(callback) {                     ❻
        console.log('discoverServicesAndCharacteristics');
        sensorTag.discoverServicesAndCharacteristics(callback);
      },
      function(callback) {
        console.log('enableIrTemperature');    ❼
        sensorTag.enableIrTemperature(callback);
      },
      function(callback) {
        setTimeout(callback, 100);             ❽
      },
      function(callback) {
        console.log('readIrTemperature');      ❾
        sensorTag.readIrTemperature(
          function(objectTemperature, ambientTemperature) {
            console.log('\tobject  temperature = %d °C',
                objectTemperature.toFixed(1));
            console.log('\tambient temperature = %d °C',
                ambientTemperature.toFixed(1));
            callback();
        });

        sensorTag.on('irTemperatureChange',    ❿
          function(objectTemperature, ambientTemperature) {
            console.log('\tobject  temperature = %d °C',
                objectTemperature.toFixed(1));
            console.log('\tambient temperature = %d °C\n',
                ambientTemperature.toFixed(1));
        });
```

```
      sensorTag.notifyIrTemperature(function() {
        console.log('notifyIrTemperature');
      });
    },
    // function(callback) {
    //   console.log('disableIrTemperature');     ⓫
    //   sensorTag.disableIrTemperature(callback);
    // },

    function(callback) {
      console.log('readSimpleRead');              ⓬
      sensorTag.on('simpleKeyChange', function(left, right) {
        console.log('left: ' + left + ' right: ' + right);
        if (left && right) {
          sensorTag.notifySimpleKey(callback);    ⓭
        }
      });

      sensorTag.notifySimpleKey(function() {      ⓮
      });
    },
    function(callback) {
      console.log('disconnect');
      sensorTag.disconnect(callback);             ⓯
    }
  ]
 );
});


// The MIT License (MIT)

// Copyright (c) 2013 Sandeep Mistry

// Permission is hereby granted, free of charge, to any person obtaining a copy of
// this software and associated documentation files (the "Software"), to deal in
// the Software without restriction, including without limitation the rights to
// use, copy, modify, merge, publish, distribute, sublicense, and/or sell copies of
// the Software, and to permit persons to whom the Software is furnished to do so,
// subject to the following conditions:

// The above copyright notice and this permission notice shall be included in all
// copies or substantial portions of the Software.

// THE SOFTWARE IS PROVIDED "AS IS", WITHOUT WARRANTY OF ANY KIND, EXPRESS OR
// IMPLIED, INCLUDING BUT NOT LIMITED TO THE WARRANTIES OF MERCHANTABILITY, FITNESS
// FOR A PARTICULAR PURPOSE AND NONINFRINGEMENT. IN NO EVENT SHALL THE AUTHORS OR
// COPYRIGHT HOLDERS BE LIABLE FOR ANY CLAIM, DAMAGES OR OTHER LIABILITY, WHETHER
// IN AN ACTION OF CONTRACT, TORT OR OTHERWISE, ARISING FROM, OUT OF OR IN
// CONNECTION WITH THE SOFTWARE OR THE USE OR OTHER DEALINGS IN THE SOFTWARE.
```

❶   Read in the various packages that are needed.

❷ `SensorTag.discover` checks what SensorTags are out there. When found, it calls the inline function that follows.

❸ This function is called when the SensorTag is disconnected.

❹ Normally JavaScript does everything synchronously. Here, we want to do the following asynchronously—that is, step-by-step, one after the other. We are passing an array to `async.series()`, which contains the functions to run in the order in which they appear in the array.

❺ Connect to the SensorTag.

❻ Discover what the SensorTag can do. This is necessary before we can give it any commands.

❼ Enable temperatures. We don't get a temperature reading yet. Rather, we're instructing it to begin reading and report back when they are ready.

❽ Wait a bit for the first temperatures to be read.

❾ This specifies the function to call every time a temperature is ready. The callback is passed `objectTemperature` (what's read by the touchless IR sensors) and `ambientTemperature` (the temperature inside the SensorTag). Try putting your hand in front of the device; the `objectTemperature` should go up.

❿ Define the callback for when the temperature changes.

⓫ This commented-out code is used when you want to turn off the temperature readings.

⓬ Assign a callback to respond to the `left` and `right` button pushes.

⓭ If both buttons are pushed, pass the `callback` function to `sensorTag.notifySimpleKey()`.

⓮ `sensorTag.notifySimpleKey()` doesn't do anything in this case, but it does evaluate `callback`, allowing it to progress to the next and final state.

⓯ When we get to here, we disconnect from the SensorTag, which causes the code to exit (see **❸**).

Here's some output from the code:

```
Be sure sensorTag is on
sensorTag = {"uuid":"9059af0b8457"}
connect
discoverServicesAndCharacteristics
enableIrTemperature
readIrTemperature
    object  temperature = 2.8 °C
        ambient temperature = 0 °C
readSimpleRead
notifyIrTemperature
        object  temperature = 31.8 °C
        ambient temperature = 24.8 °C

        object  temperature = 25.9 °C
        ambient temperature = 24.8 °C

        object  temperature = 27.4 °C
        ambient temperature = 24.8 °C

        object  temperature = 32.2 °C
        ambient temperature = 24.8 °C

left: false right: true
left: true right: true
left: false right: false
disconnect
disconnected!
```

Discussion

The SensorTag can do much more than read temperatures, but we've left out many of those details to make the code more understandable. npm's sensortag page (*http://bit.ly/1MrADwk*) documents how to access the other sensors. The SensorTag node module has a nice example (*test.js*) that shows how to read all the sensors; in fact, the example in Example 2-15 is based on it. Here's where to find *test.js*:

```
bone# cd /usr/local/lib/node_modules/sensortag
bone# less test.js
```

Here, you see examples of how to read all the sensors. Look in *index.js* (in the same directory) to see all the methods that are defined.

2.12 Playing and Recording Audio

Problem

BeagleBone doesn't have audio built in, but you want to play and record files.

Solution

One approach is to buy an audio cape ("Capes" on page 318), but another, possibly cheaper approach is to buy a USB audio adapter, such as the one shown in Figure 2-22. Some adapters that I've tested are provided in "Miscellaneous" on page 318.

Figure 2-22. A USB audio dongle

Drivers for the Advanced Linux Sound Architecture (*http://bit.ly/1MrAJUR*) (ALSA) are already installed on the Bone. You can list the recording and playing devices on your Bone by using aplay and arecord, as shown in Example 2-16. BeagleBone Black has audio-out on the HDMI interface. It's listed as card 0 in Example 2-16. card 1 is my USB audio adapter's audio out.

Example 2-16. Listing the ALSA audio output and input devices on the Bone

```
bone# aplay -l
**** List of PLAYBACK Hardware Devices ****
card 0: Black [TI BeagleBone Black], device 0: HDMI nxp-hdmi-hifi-0 []
  Subdevices: 1/1
  Subdevice #0: subdevice #0
card 1: Device [C-Media USB Audio Device], device 0: USB Audio [USB Audio]
  Subdevices: 1/1
```

```
  Subdevice #0: subdevice #0

bone# arecord -l
**** List of CAPTURE Hardware Devices ****
card 1: Device [C-Media USB Audio Device], device 0: USB Audio [USB Audio]
  Subdevices: 1/1
  Subdevice #0: subdevice #0
```

In the `aplay` output shown in Example 2-16, you can see the USB adapter's audio out. By default, the Bone will send audio to the HDMI. You can change that default by creating a file in your home directory called *~/.asoundrc* and adding the code in Example 2-17 to it.

Example 2-17. Change the default audio out by putting this in ~/.asoundrc (audio.asoundrc)

```
pcm.!default {
  type plug
  slave {
    pcm "hw:1,0"
  }
}
ctl.!default {
  type hw
  card 1
}
```

You can easily play *.wav* files with `aplay`:

```
bone# aplay test.wav
```

You can play other files in other formats by installing `mplayer`:

```
bone# apt-get update
bone# apt-get install mplayer
bone# mplayer test.mp3
```

Discussion

Adding the simple USB audio adapter opens up a world of audio I/O on the Bone.

Displays and Other Outputs

3.0 Introduction

In this chapter, you will learn how to control physical hardware via BeagleBone Black's general-purpose input/output (GPIO) pins. The Bone has 65 GPIO pins that are brought out on two 46-pin headers, called P8 and P9, as shown in Figure 3-1.

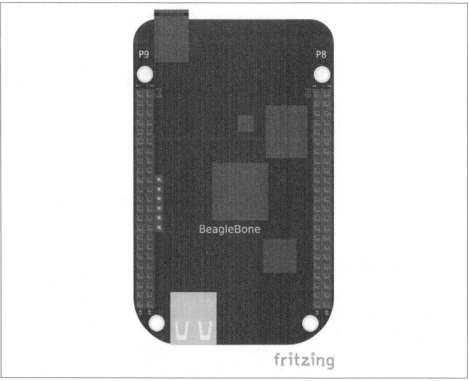

Figure 3-1. The P8 and P9 GPIO headers

The purpose of this chapter is to give simple examples that show how to use various methods of output. Most solutions require a breadboard and some jumper wires.

All these examples assume that you know how to edit a file (Recipe 1.6) and run it, either within Cloud9 integrated development environment (IDE) or from the command line (Recipe 5.3).

3.1 Toggling an Onboard LED

Problem

You want to know how to flash the four LEDs that are next to the Ethernet port on the Bone.

Solution

Locate the four onboard LEDs shown in Figure 3-2. They are labeled USR0 through USR3, but we'll refer to them as the USER LEDs.

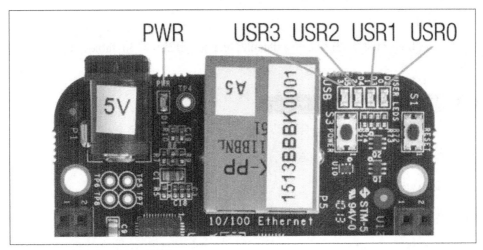

Figure 3-2. The four USER LEDs

Place the code shown in Example 3-1 in a file called *internLED.js*. You can do this using Cloud9 to edit files (as shown in Recipe 1.6) or with a more traditional editor (as shown in Recipe 5.9).

Example 3-1. Using an internal LED (internLED.js)

```
#!/usr/bin/env node
var b = require('bonescript');
var LED = 'USR0';
var state = b.HIGH;      // Initial state
b.pinMode(LED, b.OUTPUT);

setInterval(flash, 250);    // Change state every 250 ms

function flash() {
    b.digitalWrite(LED, state);
    if(state === b.HIGH) {
        state = b.LOW;
    } else {
        state = b.HIGH;
    }
}
```

In the bash command window, enter the following commands:

```
bone# chmod +x internLED.js
bone# ./internLED.js
```

The USER0 LED should now be flashing.

Discussion

The four onboard USER LEDs have a default behavior, as shown in Table 3-1.

Table 3-1. Default LED behavior

LED	Pattern
USER0	Heartbeat pattern
USER1	Flashes when the microSD is being accessed
USER2	Flashes when the CPU is busy
USER3	Flashes when the onboard flash is being accessed

When you run your code, it turns off the default action and gives you full control.

3.2 Toggling an External LED

Problem

You want to connect your own external LED to the Bone.

Solution

Connect an LED to one of the GPIO pins using a series resistor to limit the current. To make this recipe, you will need:

- Breadboard and jumper wires (see "Prototyping Equipment" on page 316)
- 220 Ω to 470 Ω resistor (see "Resistors" on page 316)
- LED (see "Opto-Electronics" on page 318)

 The value of the current limiting resistor depends on the LED you are using. The Bone can drive only 4 to 6 mA, so you might need a larger resistor to keep from pulling too much current. A 330 Ω or 470 Ω resistor might be better.

Figure 3-3 shows how you can wire the LED to pin 14 of the P9 header (P9_14). Every circuit in this book (Recipe 1.5) assumes you have already wired the rightmost bus to ground (P9_1) and the next bus to the left to the 3.3 V (P9_3) pins on the header. Be sure to get the polarity right on the LED. The *short* lead always goes to ground.

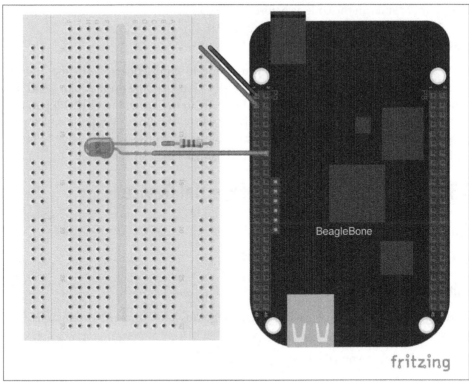

Figure 3-3. Diagram for using an external LED

After you've wired it, start Cloud9 (see Recipe 1.6) and enter the code shown in Example 3-2 in a file called *externLED.js*.

Example 3-2. Code for using an external LED (externLED.js)

```
#!/usr/bin/env node
var b = require('bonescript');
var LED = 'P9_14';
var state = b.HIGH;      // Initial state
b.pinMode(LED, b.OUTPUT);

setInterval(flash, 250);    // Change state every 250 ms

function flash() {
    b.digitalWrite(LED, state);
    if(state === b.HIGH) {
        state = b.LOW;
    } else {
        state = b.HIGH;
    }
}
```

Save your file and run the code as before (Recipe 3.1).

Discussion

LEDs are a useful and easy way for your Bone to communicate with the outside world. But be careful: LEDs operate on 1.2 to 1.4 V (depending on the color), and the Bone supplies 3.3 V. If you attach an LED directly to the Bone, the LED will pull too much current and might damage the LED, or worse, the Bone. We put a resistor in series to limit the current. 220 Ω works well, but you can use a larger value to make the LED less bright. We wouldn't recommend using a much smaller value.

3.3 Toggling a High-Voltage External Device

Problem

You want to control a device that runs at 120 V.

Solution

Working with 120 V can be tricky—even dangerous—if you aren't careful. Here's a safe way to do it.

To make this recipe, you will need:

- PowerSwitch Tail II (see "Miscellaneous" on page 318)

Figure 3-4 shows how you can wire the PowerSwitch Tail II to pin P9_14.

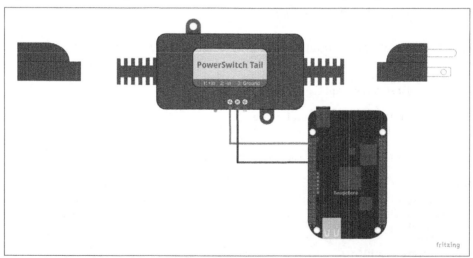

Figure 3-4. Diagram for wiring PowerSwitch Tail II

After you've wired it, because this uses the same output pin as Recipe 3.2, you can run the same code (Example 3-2).

Discussion

This wiring causes the PowerSwitch to operate in the opposite way you would likely think. When you output a HIGH to P9_14, the PowerSwitch will turn off. When you output a LOW, it turns on. Read on to see why.

The GPIO pins on the Bone can *source* 4 or 6 mA at 3.3 V, depending on the pin. This means it can supply a positive current of 4 or 6 mA. All GPIO pins can *sink* 8 mA—that is, they can absorb up to 8 mA. The PowerSwitch Tail II requires 3 to 30 mA and 3 to 12 V to turn on. P9_14 cannot *source* enough current to cleanly turn on the PowerSwitch, but it can *sink* enough; therefore, we've wired the +in input to the 3.3 V power supply on the Bone (P9_3) and the -in input to P9_14.

When P9_14 is at 3.3 V, the PowerSwitch doesn't see a voltage drop (both inputs are at 3.3 V), so it doesn't turn on.

When P9_14 drops to 0 V, there is a 3.3 V drop across the PowerSwitch, and the internal relay clicks on. Therefore, a HIGH output makes it turn off, and a LOW makes it turn on.

3.4 Fading an External LED

Problem

You want to change the brightness of an LED from the Bone.

Solution

BoneScript has an analogWrite() function that uses the Bone's pulse width modulation (PWM) hardware to produce an analog out signal. We'll use the same circuit as before (Figure 3-3) and declare the pin mode to be ANALOG_OUTPUT. Add the code in Example 3-3 to a file called *fadeLED.js* and then run it as before.

Example 3-3. Code for using an external LED (fadeLED.js)

```
#!/usr/bin/env node
var b = require('bonescript');
var LED = 'P9_14';   // Pin to use
var step = 0.02,     // Step size
    min = 0.02,      // dimmest value
    max = 1,         // brightest value
    brightness = min; // Current brightness;

b.pinMode(LED, b.ANALOG_OUTPUT, 6, 0, 0, doInterval);

function doInterval(x) {
    if(x.err) {
        console.log('x.err = ' + x.err);
        return;
    }
    setInterval(fade, 20);      // Step every 20 ms
}

function fade() {
    b.analogWrite(LED, brightness);
    brightness += step;
    if(brightness >= max || brightness <= min) {
        step = -1 * step;
    }
}
```

Discussion

The code uses setInterval() to call fade() regularly. When called, fade() will increase the brightness by step. Note that, if step is negative, the LED will actually become dimmer. If brightness is outside the min-max bounds, the step direction is changed.

You will need to press ^C (Ctrl-C) to quit the script.

You cannot use every pin for analog writes; only those that support PWM will work. Point your browser to the Headers page served by your Bone (*http://192.168.7.2/Support/bone101/#headers*) and scroll down to the PWM page to see which pins work. The table is reproduced in Figure 3-5.

P9					P8			
DGND	1	2	DGND		DGND	1	2	DGND
VDD_3K3	3	4	VDD_3V3		GPIO_38	3	4	GPIO_39
VDD_5V	5	6	VDD_5V		GPIO_34	5	6	GPIO_35
SYS_5V	7	8	SYS_5V		TIMER4	7	8	TIMER7
PWR_BUT	9	10	SYS_RESETN		TIMER5	9	10	TIMER6
GPIO_30	11	12	GPIO_60		GPIO_45	11	12	GPIO_44
GPIO_31	13	14	EHRPWM1A		EHRPWM2B	13	14	GPIO_26
GPIO_48	15	16	EHRPWM1B		GPIO_47	15	16	GPIO_46
GPIO_5	17	18	GPIO_4		GPIO_27	17	18	GPIO_65
	19	20			EHRPWM2A	19	20	GPIO_63
EHRPWM0B	21	22	EHRPWM0A		GPIO_62	21	22	GPIO_37
GPIO_49	23	24	GPIO_15		GPIO_36	23	24	GPIO_33
GPIO_117	25	26	GPIO_14		GPIO_32	25	26	GPIO_61
GPIO_115	27	28	ECAPPWM2		GPIO_86	27	28	GPIO_88
EHRPWM0B	29	30	GPIO_112		GPIO_87	29	30	GPIO_89
EHRPWM0A	31	32	VDD_ADC		GPIO_10	31	32	GPIO_11
AIN4	33	34	GNDA_ADC		GPIO_9	33	34	EHRPWM1B
AIN6	35	36	AIN5		GPIO_8	35	36	EHRPWM1A
AIN2	37	38	AIN3		GPIO_78	37	38	GPIO_79
AIN0	39	40	AIN1		GPIO_76	39	40	GPIO_77
GPIO_20	41	42	ECAPPWM0		GPIO_74	41	42	GPIO_75
DGND	43	44	DGND		GPIO_72	43	44	GPIO_73
DGND	45	46	DGND		EHRPWM2A	45	46	EHRPWM2B

Figure 3-5. Table of PWM outputs

3.5 Writing to an LED Matrix

Problem

You have an I²C-based LED matrix to interface.

Solution

There are a number of nice LED matrices that allow you to control several LEDs via one interface. This solution uses an Adafruit Bicolor 8x8 LED Square Pixel Matrix w/I²C Backpack (*http://www.adafruit.com/products/902*).

To make this recipe, you will need:

- Breadboard and jumper wires (see "Prototyping Equipment" on page 316)
- Two 4.7 kΩ resistors (see "Resistors" on page 316)
- I²C LED matrix (see "Opto-Electronics" on page 318)

The LED matrix is a 5 V device, but you can drive it from 3.3 V. Wire, as shown in Figure 3-6.

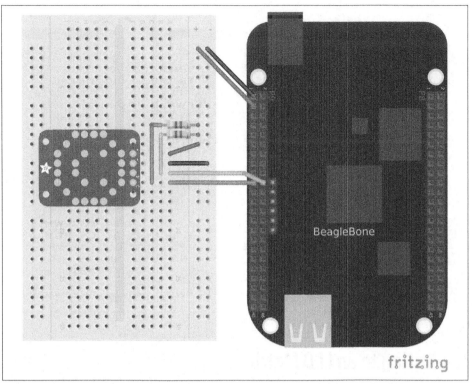

Figure 3-6. Wiring an i2c LED matrix

Recipe 2.9 shows how to use `i2cdetect` to discover the address of an I²C device. It also describes the difference between how Linux numbers the I²C buses (`0` and `1`) versus how BoneScript numbers them (`1` and `2`). Let's see how your display shows up.

> The BoneScript library convention is to use the index numbers provided in the hardware documentation. The version of the Linux kernel we use, however, begins index numbers at `0` for the first one registered and increases by 1, so these numbers might not always match. The BoneScript library attempts to hide this complication from you.

Run the `i2cdetect -y -r 1` command to discover the address of the display, as shown in Example 3-4.

Example 3-4. Using I²C command-line tools to discover the address of the display

```
bone# i2cdetect -y -r 1
     0  1  2  3  4  5  6  7  8  9  a  b  c  d  e  f
00:          -- -- -- -- -- -- -- -- -- -- -- --
10: -- -- -- -- -- -- -- -- -- -- -- -- -- -- -- --
20: -- -- -- -- -- -- -- -- -- -- -- -- -- -- -- --
30: -- -- -- -- -- -- -- -- -- -- -- -- -- -- -- --
40: -- -- -- -- -- -- -- -- -- 49 -- -- -- -- -- --
50: -- -- -- -- UU UU UU UU -- -- -- -- -- -- -- --
60: -- -- -- -- -- -- -- -- -- -- -- -- -- -- -- --
70: 70 -- -- -- -- -- -- --
```

Here, you can see a device at 0x49 and 0x70. I know I have a temperature sensor at 0x49, so the LED matrix must be at 0.70.

Add the code in Example 3-5 to a file called *matrixLEDi2c.js* and run it by using the following command:

```
bone# npm install -g sleep
bone# ./matrixLEDi2c.js
```

Example 3-5. LED matrix display (matrixLEDi2c.js)

```
#!/usr/bin/env node
// npm install -g sleep

var b = require('bonescript');
var sleep = require('sleep');
var port = '/dev/i2c-2'                              ❶
var matrix = 0x70;                                   ❷
var time = 1000000; // Delay between images in us

// The first btye is GREEN, the second is RED.       ❸
var smile =
        [0x00, 0x3c, 0x00, 0x42, 0x28, 0x89, 0x04, 0x85,
         0x04, 0x85, 0x28, 0x89, 0x00, 0x42, 0x00, 0x3c];
var frown =
        [0x3c, 0x00, 0x42, 0x00, 0x85, 0x20, 0x89, 0x00,
         0x89, 0x00, 0x85, 0x20, 0x42, 0x00, 0x3c, 0x00];
var neutral =
        [0x3c, 0x3c, 0x42, 0x42, 0xa9, 0xa9, 0x89, 0x89,
         0x89, 0x89, 0xa9, 0xa9, 0x42, 0x42, 0x3c, 0x3c];
var blank = [0, 0, 0, 0, 0, 0, 0, 0, 0, 0, 0, 0, 0, 0, 0, 0];

b.i2cOpen(port, matrix);                             ❹

b.i2cWriteByte(port,  0x21); // Start oscillator (p10)   ❺
b.i2cWriteByte(port,  0x81); // Disp on, blink off (p11)
b.i2cWriteByte(port,  0xe7); // Full brightness (page 15)
```

```
b.i2cWriteBytes(port, 0x00, frown);                          ❻
sleep.usleep(time);

b.i2cWriteBytes(port, 0x00, neutral);
sleep.usleep(time);

b.i2cWriteBytes(port, 0x00, smile);
// Fade the display
var fade;
for(fade = 0xef; fade >= 0xe0; fade--) {                     ❼
    b.i2cWriteByte(port,  fade);
    sleep.usleep(time/10);
}
for(fade = 0xe1; fade <= 0xef; fade++) {
    b.i2cWriteByte(port,  fade);
    sleep.usleep(time/10);
}
b.i2cWriteBytes(port, 0x04, [0xff]);
```

❶   This line states which bus to use. The last digit gives the BoneScript bus number.

❷   This specifies the address of the LED matrix, 0x70 in our case.

❸   This indicates which LEDs to turn on. The first byte is for the first column of
     green LEDs. In this case, all are turned off. The next byte is for the first column of
     red LEDs. The hex 0x3c number is 0b00111100 in binary. This means the first
     two red LEDs are off, the next four are on, and the last two are off. The next byte
     (0x00) says the second column of *green* LEDs are all off, the fourth byte (0x42 =
     0b01000010) says just two red LEDs are on, and so on. Declarations define four
     different patterns to display on the LED matrix, the last being all turned off.

❹   Open the I²C port.

❺   Send three commands to the matrix to get it ready to display.

❻   Now, we are ready to display the various patterns. After each pattern is displayed,
     we sleep a certain amount of time so that the pattern can be seen.

❼   Finally, send commands to the LED matrix to set the brightness. This makes the
     disply fade out and back in again.

Discussion

This setup assumes that the LED matrix can be driven from 3.3 V. See Recipe 3.6 to
drive a 5 V display.

You don't need to write the entire LED matrix with each write. For example, `b.i2cWri teBytes(port, 0x04, [0xff])`; would turn on the third columns of *green* LEDs, without writing the other columns.

3.6 Driving a 5 V Device

Problem

You have a 5 V device to drive, and the Bone has 3.3 V outputs.

Solution

If you are lucky, you might be able to drive a 5 V device from the Bone's 3.3 V output. Try it and see if it works. If not, you need a level translator.

What you will need for this recipe:

- A PCA9306 level translator (see "Integrated Circuits" on page 317)
- A 5 V power supply (if the Bone's 5 V power supply isn't enough)

The PCA9306 translates signals at 3.3 V to 5 V in both directions. It's meant to work with I²C devices that have a pull-up resistor, but it can work with anything needing translation.

Figure 3-7 shows how to wire a PCA9306 to an LED matrix. The left is the 3.3 V side and the right is the 5 V side. Notice that we are using the Bone's built-in 5 V power supply.

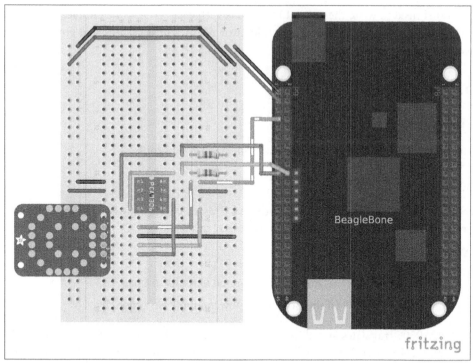

Figure 3-7. Wiring a PCA9306 level translator to an LED matrix

If your device needs more current than the Bone's 5 V power supply provides, you can wire in an external power supply.

Discussion

The 3.3 V clock and data signals from the Bone enter the PCA9306 on pins 3 and 4 (the bottom two on the left) and are translated to 5 V. Then they leave via pins 5 and 6 (lower right). These then drive the 5 V device.

3.7 Writing to a NeoPixel LED String

Problem

You have an Adafruit NeoPixel LED string (*http://www.adafruit.com/products/1138*) or Adafruit NeoPixel LED matrix (*http://www.adafruit.com/products/1487*) and want to light it up.

Solution

Wire up an Adafruit NeoPixel LED 8-by-8 matrix as shown in Figure 3-8.

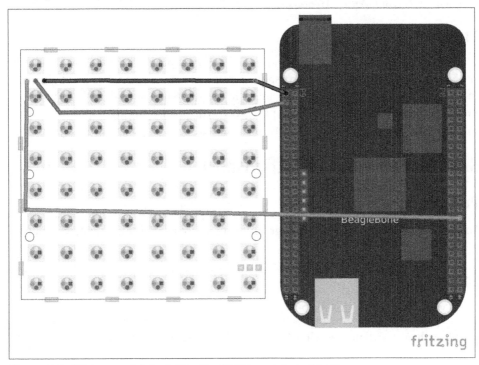

Figure 3-8. Wiring an Adafruit NeoPixel LED matrix to P8_30

Example 3-6 shows how to install LEDscape and run the LEDs.

Example 3-6. Installing and running LEDscape and OpenPixelControl (neoPixel.sh)

```
#!/bin/sh
# Here's what you do to install the neoPixel driver
# Disable the HDMI to gain access to the PRU pins
sed -i '/cape_disable=capemgr.disable_partno=BB-BONELT-HDMI,BB-BONELT-HDMIN$/ \
        s/^#//' /boot/uEnv.txt
reboot
# Clone and build the code
cd
git clone -b opc-server https://github.com/jadonk/LEDscape.git
cd LEDscape
make
cd
git clone https://github.com/jadonk/openpixelcontrol.git
# Load and configure the kernel module, pins and LEDscape daemon
config-pin overlay BB-OPC-EX
```

```
modprobe uio_pruss
./LEDscape/run-ledscape &
# Run an example Python script
./openpixelcontrol/python_clients/example.py
```

Discussion

This code utilizes the LEDscape library on the Bone's programmable real-time unit (PRU) microcontrollers. We haven't spoken much about those yet, but I will give you more ideas regarding working with them in Recipe 8.6. Using the PRUs, you can implement many different interface protocols with precise timing control, and you can continue your development on the Bone in high-level languages utilizing Linux.

3.8 Using a Nokia 5510 LCD Display

Problem

You want to display some text and graphics on a Nokia 5510 black-and-white LCD display.

Solution

What you will need for this recipe:

- Breadboard and jumper wires (see "Prototyping Equipment" on page 316)
- Nokia 5110 LCD (see "Miscellaneous" on page 318)
- 220 Ω resistor (see "Resistors" on page 316)

The Nokia 5110 LCD runs off of 3.3 V, so you can wire it directly to the Bone (Figure 3-9).

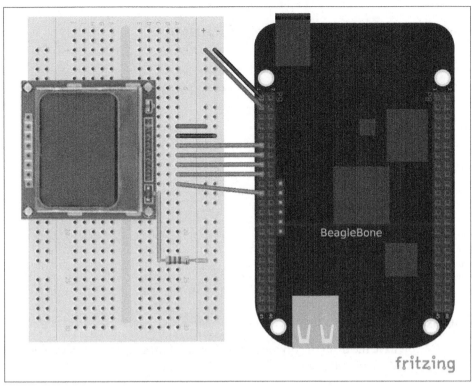

Figure 3-9. Wiring a Nokia 5110 LCD display

The drivers for the 5110 are in the Node Packaged Modules repository. Get and install them by running the following commands:

```
bone# npm install -g nokia5110
bone# cd /usr/local/lib/node_modules/nokia5110
bone# ln -s LCD_5110.js nokia5110.js
bone# cp lcdTest.js /tmp
bone# cd -
bone# mv /tmp/lcdTest.js nokia5110Test.js
```

Edit *nokia5110Test.js* so that the first few lines specify how it's wired (as shown in Example 3-7) and then run the code.

Example 3-7. Defining the Nokia 5110 pins (nokia5110.js)

```
#!/usr/bin/env node
//
// Copyright (C) 2012 - Cabin Programs, Ken Keller
//

var lcd = require('nokia5110');
var b = require('bonescript');
var timeout = 0;
var inverseIndex;

//
//  Must define the following outputs
//
lcd.PIN_SDIN = "P9_17";
lcd.PIN_SCLK = "P9_21";
lcd.PIN_SCE  = "P9_11";
lcd.PIN_DC   = "P9_15";
lcd.PIN_RESET= "P9_13";
```

Discussion

Now that you have the display working, look in `nokia5110Test.js` to see examples of how to display text and bitmaps, scroll text, and invert the display.

3.9 Making Your Bone Speak

Problem

Your Bone wants to talk.

Solution

Just install the *flite* text-to-speech program:

```
bone# apt-get install flite
```

Then add the code from Example 3-8 in a file called *speak.js* and run.

Example 3-8. A program that talks (speak.js)

```
#!/usr/bin/env node

var exec = require('child_process').exec;

function speakForSelf(phrase) {
{
```

```
    exec('flite -t "' + phrase + '"', function (error, stdout, stderr) {
        console.log(stdout);
        if(error) {
            console.log('error: ' + error);
        }
        if(stderr) {
            console.log('stderr: ' + stderr);
        }
    });
}

speakForSelf("Hello, My name is Borris. " +
    "I am a BeagleBone Black, " +
    "a true open hardware, " +
    "community-supported embedded computer for developers and hobbyists. " +
    "I am powered by a 1 Giga Hertz Sitara™ ARM® Cortex-A8 processor. " +
    "I boot Linux in under 10 seconds. " +
    "You can get started on development in " +
    "less than 5 minutes with just a single USB cable." +
    "Bark, bark!"
    );
```

See Recipe 2.12 to see how to use a USB audio dongle and set your default audio out.

Discussion

Here, speakForSelf() uses the exec() function to call the external program flite.
We pass the string to speak to speakForSelf(), and it passes the string on to flite as
a parameter to synthesize.

## CHAPTER 4

Motors

4.0 Introduction

One of the many fun things about embedded computers is that you can move physical things with motors. But there are so many different kinds of motors (*servo, stepper, DC*), so how do you select the right one?

The type of motor you use depends on the type of motion you want:

R/C or hobby servo motor
>Can be quickly positioned at various absolute angles, but some don't spin. In fact, many can turn only about 180°.

Stepper motor
>Spins and can also rotate in precise relative angles, such as turning 45°. Stepper motors come in two types: *bipolar* (which has four wires) and *unipolar* (which has five or six wires).

DC motor
>Spins either clockwise or counter-clockwise and can have the greatest speed of the three. But a DC motor can't easily be made to turn to a given angle.

When you know which type of motor to use, interfacing is easy. This chapter shows how to interface with each of these motors.

 Motors come in many sizes and types. This chapter presents some of the more popular types and shows how they can interface easily to the Bone. If you need to turn on and off a 120 V motor, consider using something like the PowerSwitch presented in Recipe 3.3.

The Bone has built-in 3.3 V and 5 V supplies, which can supply enough current to drive some small motors. Many motors, however, draw enough current that an external power supply is needed. Therefore, an external 5 V power supply is listed as optional in many of the recipes.

4.1 Controlling a Servo Motor

Problem

You want to use BeagleBone to control the absolute position of a servo motor.

Solution

We'll use the pulse width modulation (PWM) hardware of the Bone and control a servo motor with the `analogWrite()` function.

To make the recipe, you will need:

- Servo motor (see "Miscellaneous" on page 318)
- Breadboard and jumper wires (see "Prototyping Equipment" on page 316)
- 1 kΩ resistor (optional, see "Resistors" on page 316)
- 5 V power supply (optional, see "Miscellaneous" on page 318)

The 1 kΩ resistor isn't required, but it provides some protection to the general-purpose input/output (GPIO) pin in case the servo fails and draws a large current.

Wire up your servo, as shown in Figure 4-1.

There is no standard for how servo motor wires are colored. One of my servos is wired like Figure 4-1: red is 3.3 V, black is ground, and yellow is the control line. I have another servo that has red as 3.3 V and ground is brown, with the control line being orange. Generally, though, the 3.3 V is in the middle. Check the datasheet for your servo before wiring.

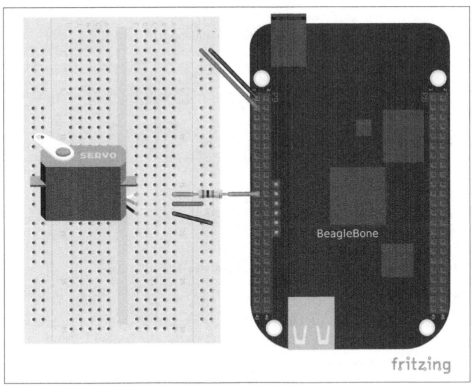

Figure 4-1. Driving a servo motor with the 3.3 V power supply

The code for controlling the servo motor is in *servoMotor.js*, shown in Example 4-1.

Example 4-1. Code for driving a servo motor (servoMotor.js)

```
#!/usr/bin/env node

// Drive a simple servo motor back and forth

var b = require('bonescript');

var motor = 'P9_21', // Pin to control servo
    freq = 50,   // Servo frequency (20 ms)
    min  = 0.8,  // Smallest angle (in ms)
    max  = 2.5,  // Largest angle (in ms)
    ms   = 250,  // How often to change position, in ms
    pos  = 1.5,  // Current position, about middle
    step = 0.1;  // Step size to next position

console.log('Hit ^C to stop');
b.pinMode(motor, b.ANALOG_OUTPUT, 6, 0, 0, doInterval);

function doInterval(x) {
    if(x.err) {
        console.log('x.err = ' + x.err);
        return;
    }
    timer = setInterval(sweep, ms);
}

move(pos);      // Start in the middle

// Sweep from min to max position and back again
function sweep() {
    pos += step;    // Take a step
    if(pos > max || pos < min) {
        step *= -1;
    }
    move(pos);
}

function move(pos) {
    var dutyCycle = pos/1000*freq;
    b.analogWrite(motor, dutyCycle, freq);
    console.log('pos = ' + pos + ' duty cycle = ' + dutyCycle);
}

process.on('SIGINT', function() {
    console.log('Got SIGINT, turning motor off');
    clearInterval(timer);            // Stop the timer
    b.analogWrite(motor, 0, freq);   // Turn motor off
});
```

Running the code causes the motor to move back and forth, progressing to successive positions between the two extremes. You will need to press ^C (Ctrl-C) to stop the script.

Discussion

Servo motors are often used in radio-controlled airplanes or cars. They generally don't spin, but rather turn only 180°, just enough to control an elevator on a wing or steer wheels on a car.

The servo is controlled by a PWM signal, which is controlled by the analogWrite() function, which sends a pulse 50 times each second (50 Hz or every 20 ms). The width of the pulse determines the position of the motor. A short pulse (1 ms, for example) sends the motor angle to zero degrees. A long pulse (2 ms, for example) sets the angle to 180 degrees.

The code in Example 4-1 defines a move() function, which moves the motor to a different angle every time it is called. The setInterval() function schedules sweep() to be called every quarter second (250 ms) to step the servo to a new position.

At the end of the code, the process.on() function detects when the user has pressed ^C (Ctrl-C), stops the timer, and turns off the motor. If you don't turn off the motor, it will keep buzzing, even if it isn't moving.

Not all GPIO pins support the PWM hardware; Recipe 3.4 discusses which pins you can use.

If your servo motor needs to run off 5 V, wire it as shown in Figure 4-2. The *banded* wire attaches the Bone's 5 V power supply (P9_7) to the power supply on the servo. We are still controlling it with a 3.3 V signal.

Here, we move the *banded* power wire from the 3.3 V power supply to the Bone's 5 V power supply on pin P9_7. If the Bone's built-in 5 V doesn't supply enough current, or if you need a higher voltage, connect your servo's power supply wire to an external power supply.

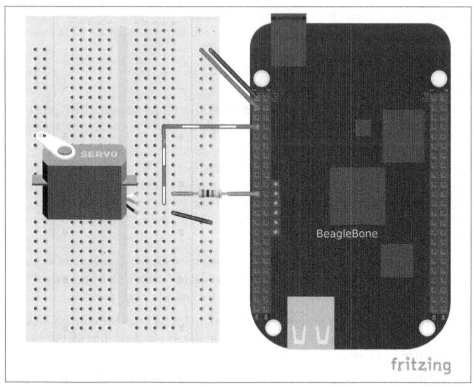

Figure 4-2. Driving a servo motor with the 5 V power supply

Of the motors presented in this chapter, the servo motor is the easiest to drive, because the control line on the servo doesn't require much current and therefore doesn't require additional components.

4.2 Controlling the Speed of a DC Motor

Problem

You have a DC motor (or a solenoid) and want a simple way to control its speed, but not the direction.

Solution

It would be nice if you could just wire the DC motor to BeagleBone Black and have it work, but it won't. Most motors require more current than the GPIO ports on the Bone can supply. Our solution is to use a transistor to control the current to the bone.

Here's what you will need:

- 3 V to 5 V DC motor
- Breadboard and jumper wires (see "Prototyping Equipment" on page 316)
- 1 kΩ resistor (see "Resistors" on page 316)
- Transistor 2N3904 (see "Transistors and Diodes" on page 317)
- Diode 1N4001 (see "Transistors and Diodes" on page 317)
- Power supply for the motor (optional)

If you are using a larger motor (more current), you will need to use a larger transistor.

Wire your breadboard as shown in Figure 4-3.

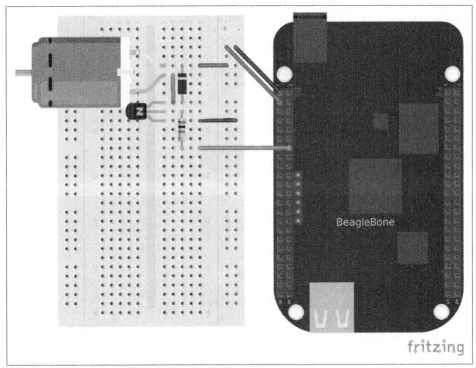

Figure 4-3. Wiring a DC motor to spin one direction

Use the code in Example 4-2 (*dcMotor.js*) to run the motor.

Example 4-2. Driving a DC motor in one direction (dcMotor.js)

```
#!/usr/bin/env node

// This is an example of driving a DC motor

var b = require('bonescript');

var motor = 'P9_16',// Pin to drive transistor
    min = 0.05,      // Slowest speed (duty cycle)
    max = 1,         // Fastest (always on)
    ms = 100,        // How often to change speed, in ms
    speed = 0.5,     // Current speed;
    step = 0.05;     // Change in speed

b.pinMode(motor, b.ANALOG_OUTPUT, 6, 0, 0, doInterval);

function doInterval(x) {
    if(x.err) {
        console.log('x.err = ' + x.err);
        return;
    }
    var timer = setInterval(sweep, ms);
}

function sweep() {
    speed += step;
    if(speed > max || speed < min) {
        step *= -1;
    }
    b.analogWrite(motor, speed);
    console.log('speed = ' + speed);
}

process.on('SIGINT', function() {
    console.log('Got SIGINT, turning motor off');
    clearInterval(timer);          // Stop the timer
    b.analogWrite(motor, 0);       // Turn motor off
});
```

Discussion

This is actually the same code that you used to drive the servo motor (Recipe 4.1). The only difference is how the parameters are set. We are now using the PWM hardware to control the speed of the motor rather than the position. If the duty cycle is 1, the voltage to the motor is on all the time, and it runs at its fastest. If the duty cycle is 0.5, the voltage is on half the time, so the motor runs slower. A duty cycle of 0 stops the motor.

You can use this same setup to drive a solenoid. After all, a solenoid is a DC motor that goes back and forth rather than spinning. Generally, you would drive a solenoid with an on or off signal, as opposed to using a PWM signal.

At the end of the code, the process.on() function detects when the user has pressed ^C (Ctrl-C), stops the timer, and turns off the motor. If you don't turn off the motor, it will continue to spin.

See Also

How do you change the direction of the motor? See Recipe 4.3.

4.3 Controlling the Speed and Direction of a DC Motor

Problem

You would like your DC motor to go forward and backward.

Solution

Use an H-bridge to switch the terminals on the motor so that it will run both backward and forward. We'll use the *L293D*: a common, single-chip H-bridge.

Here's what you will need:

- 3 V to 5 V motor (see "Miscellaneous" on page 318)
- Breadboard and jumper wires (see "Prototyping Equipment" on page 316)
- L293D H-Bridge IC (see "Integrated Circuits" on page 317)
- Power supply for the motor (optional)

Lay out your breadboard as shown in Figure 4-4. Ensure that the L293D is positioned correctly. There is a notch on one end that should be pointed up.

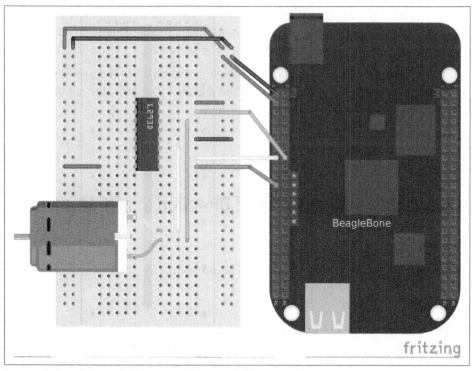

Figure 4-4. Driving a DC motor with an H-bridge

The code in Example 4-3 (*h-bridgeMotor.js*) looks much like the code for driving the DC motor with a transistor (Example 4-2). The additional code specifies which direction to spin the motor.

Example 4-3. Code for driving a DC motor with an H-bridge (h-bridgeMotor.js)

```
#!/usr/bin/env node

// This example uses an H-bridge to drive a DC motor in two directions

var b = require('bonescript');

var enable = 'P9_21';    // Pin to use for PWM speed control
    in1    = 'P9_15',
    in2    = 'P9_16',
    step = 0.05,    // Change in speed
    min  = 0.05,    // Min duty cycle
    max  = 1.0,     // Max duty cycle
    ms   = 100,     // Update time, in ms
    speed = min;    // Current speed;

b.pinMode(enable, b.ANALOG_OUTPUT, 6, 0, 0, doInterval);
```

```
b.pinMode(in1, b.OUTPUT);
b.pinMode(in2, b.OUTPUT);

function doInterval(x) {
    if(x.err) {
        console.log('x.err = ' + x.err);
        return;
    }
    timer = setInterval(sweep, ms);
}

clockwise();          // Start by going clockwise

function sweep() {
    speed += step;
    if(speed > max || speed < min) {
        step *= -1;
        step>0 ? clockwise() : counterClockwise();
    }
    b.analogWrite(enable, speed);
    console.log('speed = ' + speed);
}

function clockwise() {
    b.digitalWrite(in1, b.HIGH);
    b.digitalWrite(in2, b.LOW);
}

function counterClockwise() {
    b.digitalWrite(in1, b.LOW);
    b.digitalWrite(in2, b.HIGH);
}

process.on('SIGINT', function() {
    console.log('Got SIGINT, turning motor off');
    clearInterval(timer);          // Stop the timer
    b.analogWrite(enable, 0);      // Turn motor off
});
```

Discussion

The H-bridge provides a simple way to switch the leads on a motor so that it will reverse directions. Figure 4-5 shows how an H-bridge works.

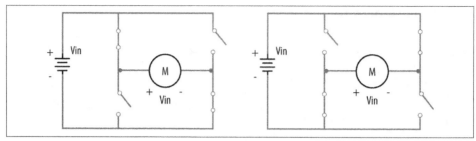

Figure 4-5. H-bridge schematic

The four switches connected to the motor (the big *M* in the center of Figure 4-5) are the H-bridge. In the left diagram, two switches are closed, connecting the left terminal of the motor to the plus voltage and the right terminal to the ground. This makes the motor rotate in one direction. The diagram on the right shows the other two switches closed, and the plus and grounds are connected to the opposite terminals, making the motor spin the other way.

In the code, in1 and in2 are used to control each pair of switches. (The L293D has four sets of these switches, but we are using only two pairs.) So, to turn clockwise, set in1 HIGH and in2 LOW. To go counter-clockwise, set them the opposite way:

```
function clockwise() {
    b.digitalWrite(in1, b.HIGH);
    b.digitalWrite(in2, b.LOW);
}

function counterClockwise() {
    b.digitalWrite(in1, b.LOW);
    b.digitalWrite(in2, b.HIGH);
}
```

The L293D also has an *enable* pin (pin 9, on the lower right), which is used to control the speed of the motor by using a PWM signal. (See Recipe 4.2 for details on how this works.)

At the end of the code, the process.on() function at the end of the code detects when the user has pressed ^C (Ctrl-C), stops the timer, and turns off the motor. If you don't turn off the motor, it will continue to spin.

> The previous example uses a motor that works with 3.3 V. What if you have a 5 V motor? The *banded* wire (running from P9_7 to pin 8 of the L293D) in Figure 4-6 attaches the L293D to the Bone's 5 V power supply. This will work if your motor doesn't draw too much current.
>
> If 5 V isn't enough voltage, or if the Bone can't supply the current needed, Figure 4-7 shows how to use an external power supply.

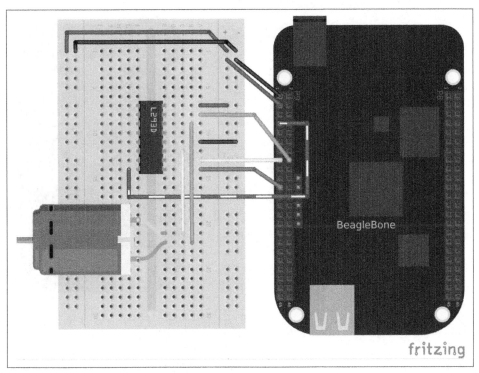

Figure 4-6. Driving a 5 V DC motor with an H-bridge

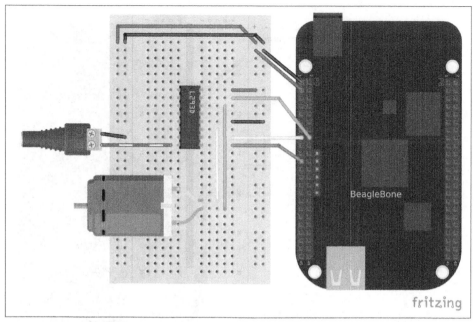

Figure 4-7. Driving a DC motor with an H-bridge and external power supply

The H-bridge provides an easy way to reverse the direction of a DC motor, and attaching the *enable* to a PWM signal allows you to control the speed.

4.4 Driving a Bipolar Stepper Motor

Problem

You want to drive a stepper motor that has four wires.

Solution

Use an L293D H-bridge. The bipolar stepper motor requires us to reverse the coils, so we need to use an H-bridge.

Here's what you will need:

- Breadboard and jumper wires (see "Prototyping Equipment" on page 316)
- 3 V to 5 V bipolar stepper motor (see "Miscellaneous" on page 318)
- L293D H-Bridge IC (see "Integrated Circuits" on page 317)

Wire as shown in Figure 4-8.

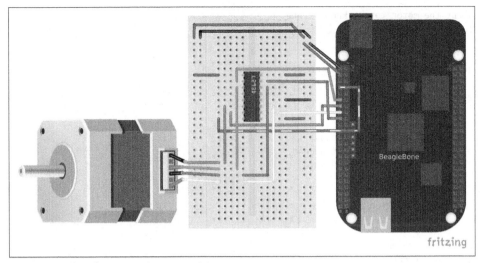

Figure 4-8. Bipolar stepper motor wiring

Use the code in Example 4-4 (*bipolarStepperMotor.js*) to drive the motor.

Example 4-4. Driving a bipolar stepper motor (bipolarStepperMotor.js)

```
#!/usr/bin/env node
var b = require('bonescript');

// Motor is attached here
var controller = ["P9_11", "P9_13", "P9_15", "P9_17"];
var states = [[1,0,0,0], [0,1,0,0], [0,0,1,0], [0,0,0,1]];
var statesHiTorque = [[1,1,0,0], [0,1,1,0], [0,0,1,1], [1,0,0,1]];
var statesHalfStep = [[1,0,0,0], [1,1,0,0], [0,1,0,0], [0,1,1,0],
                      [0,0,1,0], [0,0,1,1], [0,0,0,1], [1,0,0,1]];

var curState = 0;    // Current state
var ms = 100,        // Time between steps, in ms
    max = 22,        // Number of steps to turn before turning around
    min = 0;         // Minimum step to turn back around on

var CW  = 1,         // Clockwise
    CCW = -1,
    pos = 0,         // current position and direction
    direction = CW;

// Initialize motor control pins to be OUTPUTs
var i;
for(i=0; i<controller.length; i++) {
    b.pinMode(controller[i], b.OUTPUT);
}
```

```
// Put the motor into a known state
updateState(states[0]);
rotate(direction);

var timer = setInterval(move, ms);

// Rotate back and forth once
function move() {
    pos += direction;
    console.log("pos: " + pos);
    // Switch directions if at end.
    if (pos >= max || pos <= min) {
        direction *= -1;
    }
    rotate(direction);
}

// This is the general rotate
function rotate(direction) {
        // console.log("rotate(%d)", direction);
    // Rotate the state acording to the direction of rotation
        curState +=  direction;
        if(curState >= states.length) {
            curState = 0;
        } else if(curState<0) {
                curState = states.length-1;
        }
        updateState(states[curState]);
}

// Write the current input state to the controller
function updateState(state) {
    console.log("state: " + state);
        for (i=0; i<controller.length; i++) {
                b.digitalWrite(controller[i], state[i]);
        }
}

process.on('exit', function() {
    updateState([0,0,0,0]);    // Turn motor off
});
```

When you run the code, the stepper motor will rotate back and forth.

Discussion

Stepper motors are designed to rotate in discrete steps. They operate by turning on one coil (coil 1, for example) in one direction, then turning off coil 1 and turning on coil 2, then turning on 1 in the reverse direction, then 2 in reverse, and then back to 1 forward, as shown in Figure 4-9.

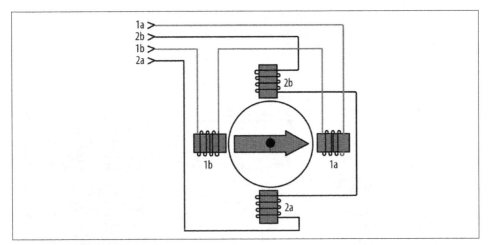

Figure 4-9. Model of a bipolar stepper motor

In Example 4-4 (*bipolarStepperMotor.js*), the `states` variable is an array of the four different states listed in the previous paragraph. That is, `[1,0,0,0]` instructs to turn on coil 1 in the forward direction. After it's been on for a while, you move to the next state, `[0,1,0,0]`. This turns off coil 1 and turns on coil 2. Then you go to `[0,0,1,0]`, which turns on coil 1 in reverse, and so on. The H-bridge (see Recipe 4.3) takes care of reversing the coil:

```
var states = [[1,0,0,0], [0,1,0,0], [0,0,1,0], [0,0,0,1]];
```

Whenever you step, you just index the next element in the array. If you want to go backward, index the previous element.

There are two other state sequences given in the code, but they aren't ever used. `statesHiTorque` is like the original `states`, except that it has both magnets on at the same time, giving a stronger pull on the motor. The `statesHalfStep` alternates between having one and two coils on simultaneously to give finer control in the step size:

```
var statesHiTorque = [[1,1,0,0], [0,1,1,0], [0,0,1,1], [1,0,0,1]];
var statesHalfStep = [[1,0,0,0], [1,1,0,0], [0,1,0,0], [0,1,1,0],
                      [0,0,1,0], [0,0,1,1], [0,0,0,1], [1,0,0,1]];
```

You can easily try either of these by assigning them to the `states` variable.

Notice the *banded* wire running from pin 8 (lower left) of the L293D to `P9_7` in Figure 4-8. This is connecting the 5 V power supply on the Bone to the L293D. The Bone outputs 3.3 V on its GPIO pins, but the stepper motor I am using really needs 5 V. Connecting the 5 V in this way provides 5 V out of the L293D to the stepper motor.

If your stepper motor requires a higher voltage (up to 12 V), you can wire an external power supply, as shown in Figure 4-10.

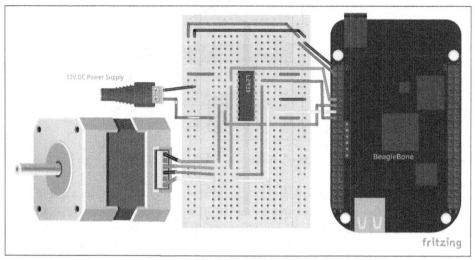

Figure 4-10. Model of a bipolar stepper motor

The stepper motor requires a more complex setup than the servo motor, but it has the advantage of being able to spin, which the servo can't do.

4.5 Driving a Unipolar Stepper Motor

Problem

You want to drive a stepper motor that has five or six wires.

Solution

If your stepper motor has five or six wires, it's a *unipolar* stepper and is wired differently than the bipolar. Here, we'll use a *ULN2003 Darlington Transistor Array IC* to drive the motor.

Here's what you will need:

- Breadboard and jumper wires (see "Prototyping Equipment" on page 316)
- 3 V to 5 V unipolar stepper motor (see "Miscellaneous" on page 318)
- ULN2003 Darlington Transistor Array IC (see "Integrated Circuits" on page 317)

Wire, as shown in Figure 4-11.

The IC in Figure 4-11 is illustrated upside down from the way it is usually displayed. That is, the notch for pin 1 is on the bottom. This made drawing the diagram much cleaner.

Also, notice the *banded* wire running the P9_7 (5 V) to the UL2003A. The stepper motor I'm using runs better at 5 V, so I'm using the Bone's 5 V power supply. The signal coming from the GPIO pins is 3.3 V, but the U2003A will step them up to 5 V to drive the motor.

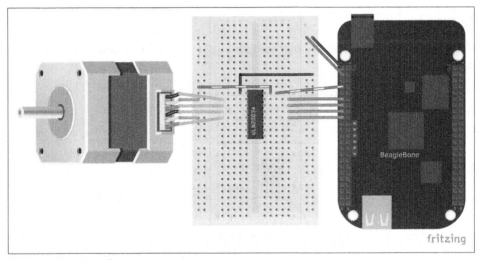

Figure 4-11. Unipolar stepper motor wiring

The code for driving the motor is in *unipolarStepperMotor.js*; however, it is almost identical to the bipolar stepper code (Example 4-4), so Example 4-5 shows only the lines that you need to change.

Example 4-5. Changes to bipolar code to drive a unipolar stepper motor (unipolarStepperMotor.diff)

```
var controller = ["P9_11", "P9_13", "P9_15", "P9_17"];
var states = [[1,1,0,0], [0,1,1,0], [0,0,1,1], [1,0,0,1]];
var curState = 0;    // Current state
var ms = 100,        // Time between steps, in ms
    max = 22,        // Number of steps to turn before turning around
```

The code in this example makes the following changes:

- `controller` is attached to the *even*-numbered pins on the P9 header rather than the *odd* that the bipolar stepper used. (Doing this allows you to run both types of stepper motors at the same time!)

- The `states` are different. Here, we have two pins high at a time.

- The time between steps (`ms`) is shorter, and the number of steps per direction (`max`) is bigger. The unipolar stepper I'm using has many more steps per rotation, so I need more steps to make it go around.

Discussion

Figure 4-12 shows how the unipolar stepper motor is wired a bit differently than the bipolar.

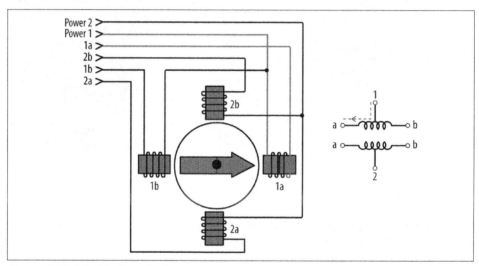

Figure 4-12. Model of a unipolar stepper motor

Each coil is split, so you can access the two sides separately. The difference between the five- and six-wire steppers is that the five-wire steppers have the two power lines connected to each other. The different wiring means a different energizing sequence, as seen in the different `states` array.

 If your stepper requires more voltage, or more current than what the Bone can supply, you can replace the *banded* wire with an external power supply, similar to the bipolar stepper example (Figure 4-10).

CHAPTER 5

Beyond the Basics

5.0 Introduction

In Chapter 1, you learned how to set up BeagleBone Black, and Chapter 2, Chapter 3, and Chapter 4 showed how to interface to the physical world. Chapter 6 through the remainder of the book moves into some more exciting advanced topics, and this chapter gets you ready for them.

The recipes in this chapter assume that you are running Linux on your host computer (Recipe 5.2) and are comfortable with using Linux. We continue to assume that you are logged in as root on your Bone.

5.1 Running Your Bone Standalone

Problem

You want to use BeagleBone Black as a desktop computer with keyboard, mouse, and an HDMI display.

Solution

The Bone comes with USB and a microHDMI output. All you need to do is connect your keyboard, mouse, and HDMI display to it.

To make this recipe, you will need:

- Standard HDMI cable and female HDMI-to-male microHDMI adapter (see "Miscellaneous" on page 318), or
- MicroHDMI-to-HDMI adapter cable (see "Miscellaneous" on page 318)

- HDMI monitor (see "Miscellaneous" on page 318)
- USB keyboard and mouse
- Powered USB hub (see "Miscellaneous" on page 318)

 The microHDMI adapter is nice because it allows you to use a regular HDMI cable with the Bone. However, it will block other ports and can damage the Bone if you aren't careful. The microHDMI-to-HDMI cable won't have these problems.

 You can also use an HDMI-to-DVI cable ("Miscellaneous" on page 318) and use your Bone with a DVI-D display.

The adapter looks something like Figure 5-1.

Figure 5-1. Female HDMI-to-male microHDMI adapter

Plug the small end into the microHDMI input on the Bone and plug your HDMI cable into the other end of the adapter and your monitor. If nothing displays on your Bone, reboot.

If nothing appears after the reboot, edit the */boot/uEnv.txt* file. Search for the line containing ##Disable HDMI and add the following lines after it:

```
##Disable HDMI
#cape_disable=capemgr.disable_partno=BB-BONELT-HDMI,BB-BONELT-HDMIN
cape_disable=capemgr.disable_partno=BB-BONELT-HDMI
```

Then reboot.

The */boot/uEnv.txt* file contains a number of configuration commands that are executed at boot time. The # character is used to add comments; that is, everything to the right of a # is ignored by the Bone and is assumed to be for humans to read. In the previous example, ##Disable HDMI is a comment that informs us the next line(s) are for disabling the HDMI. Two cape.disable commands follow. The first one is commented-out and won't be executed by the Bone.

Why not just remove the first cape.disable? Later, you might decide you need more general-purpose input/output (GPIO) pins and don't need the HDMI display. If so, just remove the # from for first cape.disble. If you had completely removed the line earlier, you would have to look up the details somewhere to re-create it.

When in doubt, comment-out; don't delete.

If you want to re-enable the HDMI audio, just comment-out the line you added.

The Bone has only one USB port, so you will need to get either a keyboard with a USB hub (see "Miscellaneous" on page 318) or a powered USB hub. Plug the USB hub into the Bone and then plug your keyboard and mouse in to the hub. You now have a Beagle workstation; no host computer is needed.

The powered hub is recommended because USB can supply only 500 mA, and you'll want to plug many things into the Bone.

Discussion

This recipe disables the HDMI audio, which allows the Bone to try other resolutions. If this fails, see BeagleBoneBlack HDMI (*http://bit.ly/1GEPcOH*) for how to force the Bone's resolution to match your monitor.

5.2 Selecting an OS for Your Development Host Computer

Problem

Your project needs a host computer, and you need to select an operating system (OS) for it.

Solution

For projects that require a host computer, we assume that you are running Linux Ubuntu 14.04 LTS (*http://bit.ly/1wXOwkw*). You can be running either a native installation, via a virtual machine such as VirtualBox (*https://www.virtualbox.org/*), or in the cloud (Microsoft Azure (*http://bit.ly/1wu04LI*) or Amazon Elastic Compute Cloud (*http://aws.amazon.com/ec2/*) [EC2], for example).

Discussion

Most of the projects in this book run on the Bone using JavaScript and the BoneScript API, and which OS is running on the host doesn't matter, as long as it has a browser (either Google Chrome (*http://bit.ly/1DAx5Zp*) or Firefox (*http://mzl.la/1BwDS2K*) work). In fact, if you attach an HDMI monitor, keyboard, and mouse (Recipe 5.1), you don't need a host computer at all.

> Other versions of Ubuntu should work. Even other Linux distributions *should* work; *this book has been tested only on Ubuntu 14.04.* Because OS X is built on Unix, the projects should also work on a Mac.

If you are running Microsoft Windows (*http://www.microsoft.com/*), many recipes will work via the browser (but maybe not Internet Explorer (*http://bit.ly/1MrBxJj*)). But if you are doing development work, you will need to run Linux. One option is to set up your Windows-based computer to dual boot (*http://bit.ly/1GrF34e*). When you power up, you'll need to choose either Ubuntu or Windows to boot up.

Another option is to run Ubuntu in a virtual machine (such as VirtualBox [*https://www.virtualbox.org/*]) on your Windows computer. In this configuration, you are running Windows, but Ubuntu runs in an application within Windows. You have access to Windows even while running Ubuntu.

Yet another option is to use one of the many free cloud services, such as Azure (*http://azure.microsoft.com/en-us/*) or EC2 (*http://aws.amazon.com/ec2/*). These are both quick and easy ways to get access to Ubuntu 14.04 without installing it on a local computer.

I've provided tutorials online that show how to set up Azure (*http://bit.ly/1E5wkm0*) and EC2 (*http://bit.ly/1NJSuSL*).

5.3 Getting to the Command Shell via SSH

Problem

You want to connect to the command shell of a remote Bone from your host computer.

Solution

Recipe 1.7 shows how to run shell commands in the Cloud9 bash tab. However, the Bone has Secure Shell (SSH) enabled right out of the box, so you can easily connect by using the following command to log in as superuser `root`, (note the # at the end of the prompt):

```
host$ ssh root@192.168.7.2
Warning: Permanently added 'bone,192.168.7.2' (ECDSA) to the list of known hosts.
Last login: Mon Dec 22 07:53:06 2014 from yoder-linux.local
bone#
```

Or you could log in as a normal user, `debian` (note the $ at the end of the prompt):

```
host$ ssh debian@192.168.7.2
Warning: Permanently added 'bone,192.168.7.2' (ECDSA) to the list of known hosts.
debian@bone's password:
Last login: Fri Dec  5 07:57:06 2014 from 192.168.7.1
bone$
```

`root` has no password. It's best to change both passwords:

```
bone$ passwd
Changing password for debian.
(current) UNIX password:
Enter new UNIX password:
Retype new UNIX password:
passwd: password updated successfully
```

Discussion

Sometimes, when connecting via SSH, you will get an error like this:

```
host$ ssh root@192.168.7.2
@@@@@@@@@@@@@@@@@@@@@@@@@@@@@@@@@@@@@@@@@@@@@@@@@@@@@@@@@@@@@@@@@@@
@      WARNING: REMOTE HOST IDENTIFICATION HAS CHANGED!        @
@@@@@@@@@@@@@@@@@@@@@@@@@@@@@@@@@@@@@@@@@@@@@@@@@@@@@@@@@@@@@@@@@@@
IT IS POSSIBLE THAT SOMEONE IS DOING SOMETHING NASTY!
Someone could be eavesdropping on you right now (man-in-the-middle attack)!
It is also possible that a host key has just been changed.
The fingerprint for the ECDSA key sent by the remote host is
1b:dc:bc:c6:8f:87:f7:de:30:97:a4:7b:84:9d:84:ad.
Please contact your system administrator.
Add correct host key in /home/yoder/.ssh/known_hosts to get rid of this message.
Offending ECDSA key in /home/yoder/.ssh/known_hosts:3
  remove with: ssh-keygen -f "/home/yoder/.ssh/known_hosts" -R 192.168.7.2
ECDSA host key for 192.168.7.2 has changed, you have requested strict checking.
Host key verification failed.
```

A quick fix is to follow the directions in the error:

```
host$ ssh-keygen -f "/home/yoder/.ssh/known_hosts" -R 192.168.7.2
```

A longer-term solution is to create an SSH configuration file. Add the following code to a file called *~/.ssh/config*:

```
Host 192.168.7.2
   User root
   UserKnownHostsFile /dev/null
   StrictHostKeyChecking no
```

The UserKnownHostsFile parameter says to check /dev/null for a list of hosts it knows, but /dev/null is a file that is always empty, so SSH will assume it doesn't know any hosts from before. The StrictHostKeyChecking no setting tells SSH to automatically add new hosts without asking.

The User root line solves a different problem. I generally log in to my Bone as root; this line makes SSH default to root, so I can enter the SSH without the root@, and it will log me in as root:

```
host$ ssh 192.168.7.2
root@192.168.7.2's password:
bone#
```

You can have a password on your Bone, but not have to type it in everytime you SSH to the Bone. From your host, do this:

```
host$ ssh-copy-id root@192.168.7.2
The authenticity of host '192.168.7.2 (192.168.7.2)'
    can't be established.
ECDSA key fingerprint is
    54:ce:02:e5:83:3f:01:b3:bc:fd:43:fe:08:d4:97:ff.
Are you sure you want to continue connecting
    (yes/no)? yes
Warning: Permanently added '192.168.7.2' (ECDSA)
    to the list of known hosts.
Now try logging into the machine, with
    "ssh 'root@192.168.7.2'", and check in:

    ~/.ssh/authorized_keys

to make sure we haven't added extra keys that you weren't
expecting.
```

This copies some keys to the Bone that tell it to allow you to log in from your host without using a password.

SSH is a very common way to connect between computers. After it's working, it opens up many other useful commands that are based on it. For example, you can copy files from one computer to another by using the scp command:

```
host$ scp fileOnHost root@192.168.7.2:fileOnBone
```

This copies the *fileOnHost* file to the Bone and puts it in the */root/fileOnBone* file.

If your host computer supports X-Windows (Linux and the Mac's OS X do), you can run your graphical program directly from the command line. When connecting to your Bone via SSH, use the -X flag. This instructs SSH to forward the graphical commands back to your host.

```
host$ ssh -X debian@192.168.7.2
bone# lxterminal
```

An lxterminal window will show up on your host.

5.4 Getting to the Command Shell via the Virtual Serial Port

Problem

You want to connect to the command shell of a remote Bone from your host computer without using SSH.

Solution

Sometimes, you can't connect to the Bone via SSH, but you have a network working over USB to the Bone. There is a way to access the command line to fix things without requiring extra hardware. (Recipe 5.5 shows a way that works even if you don't have a network working over USB, but it requires a special serial-to-USB cable.)

First, check to ensure that the serial port is there. On the host computer, run the following command:

```
host$ ls -ls /dev/ttyACM0
0 crw-rw---- 1 root dialout 166, 0 Jun 19 11:47 /dev/ttyACM0
```

/dev/ttyACM0 is a serial port on your host computer that the Bone creates when it boots up. The letters `crw-rw----` show that you can't access it as a normal user. However, you *can* access it if you are part of `dialout` group. Just add yourself to the group:

```
host$ sudo adduser $USER dialout
```

You have to run `adduser` only once. Your host computer will remember the next time you boot up. Now, install and run the `screen` command:

```
host$ sudo apt-get install screen
host$ screen /dev/ttyACM0 115200
Debian GNU/Linux 7 beaglebone ttyGS0

default username:password is [debian:temppwd]

Support/FAQ: http://elinux.org/Beagleboard:BeagleBoneBlack_Debian

The IP Address for usb0 is: 192.168.7.2
beaglebone login:
```

The `/dev/ttyACM0` parameter specifies which serial port to connect to, and `115200` tells the speed of the connection. In this case, it's 115,200 bits per second.

Discussion

You can now run your favorite commands. For help, press ^A? (that is, Ctrl-A, and then ?):

```
              Screen key bindings, page 1 of 2.

              Command key:  ^A    Literal ^A:  a

  break       ^B b       license      ,            removebuf   =
  clear       C          lockscreen   ^X x         reset       Z
  colon       :          log          H            screen      ^C c
  copy        ^[ [       login        L            select      '
  detach      ^D d       meta         a            silence     _
  digraph     ^V         monitor      M            split       S
  displays    *          next         ^@ ^N sp n   suspend     ^Z z
```

```
dumptermcap  .          number      N         time       ^T t
fit          F          only        Q         title      A
flow         ^F f       other       ^A        vbell      ^G
focus        ^I         pow_break   B         version    v
hardcopy     h          pow_detach  D         width      W
help         ?          prev        ^H ^P p ^? windows   ^W w
history      { }        quit        \         wrap       ^R r
kill         K k        redisplay   ^L l      xoff       ^S s
lastmsg      ^M m       remove      X         xon        ^Q q

             [Press Space for next page; Return to end.]
```

Pressing ^A\ (Ctrl-A and then \) quits screen.

This is a handy way to get access to your Bone when the normal way (SSH) doesn't work. However, you can't pass graphics or use commands such as scp.

5.5 Viewing and Debugging the Kernel and u-boot Messages at Boot Time

Problem

You want to see the messages that are logged by BeagleBone Black come to life.

Solution

There is no network in place when the Bone first boots up, so Recipe 5.3 and Recipe 5.4 won't work. This recipe uses some extra hardware (FTDI cable) to attach to the Bone's console serial port.

To make this recipe, you will need:

- 3.3 V FTDI cable (see "Miscellaneous" on page 318)

Be sure to get a 3.3 V FTDI cable (shown in Figure 5-2), because the 5 V cables won't work.

The Bone's Serial Debug J1 connector has Pin 1 connected to ground, Pin 4 to receive, and Pin 5 to transmit. The other pins are not attached.

Figure 5-2. FTDI cable

Look for a small triangle at the end of the FTDI cable (Figure 5-3). It's often connected to the black wire.

Figure 5-3. FTDI connector

Next, look for the FTDI pins of the Bone (labeled J1 on the Bone), shown in Figure 5-4. They are next to the P9 header and begin near pin 20. There is a white dot near P9_20.

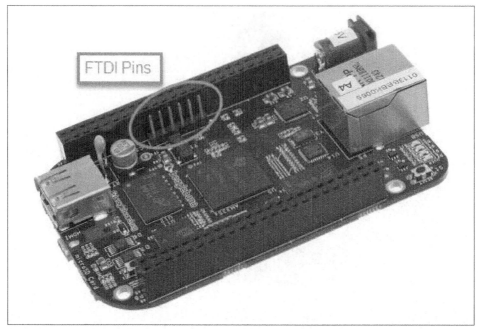

Figure 5-4. FTDI pins for the FTDI connector [1]

Plug the FTDI connector into the FTDI pins, being sure to connect the *triangle* pin on the connector to the *white dot* pin of the FTDI connector.

Now, run the following commands on your host computer:

```
host$ ls -ls /dev/ttyUSB0
0 crw-rw---- 1 root dialout 188, 0 Jun 19 12:43 /dev/ttyUSB0
host$ sudo adduser $USER dialout
host$ screen /dev/ttyUSB0 115200
Debian GNU/Linux 7 beaglebone ttyO0

default username:password is [debian:temppwd]

Support/FAQ: http://elinux.org/Beagleboard:BeagleBoneBlack_Debian

The IP Address for usb0 is: 192.168.7.2
beaglebone login:
```

---

1 *Figure 5-4 was originally posted by Jason Kridner at http://beagleboard.org/media under a Creative Commons Attribution-ShareAlike 3.0 Unported License (http://creativecommons.org/licenses/by-sa/3.0/).*

Your screen might initially be black. Press Enter a couple times to see the login prompt.

Discussion

This is the same `screen` command as in Recipe 5.4, just using a different port.

 To see and save the boot loader and Linux boot-up messages, press ^AH (Ctrl-A and then H) to start logging to a file on your host computer and then reboot your Bone.

Now you see what goes on before the kernel starts:

```
bone# reboot
Broadcast message from root@beaglebone (tty00) (Thu Jun 19 15:04:40 20
The system is going down for reboot NOW!
...
[  368.405404] Restarting system.

U-Boot SPL 2013.04-dirty (Jul 10 2013 - 14:02:53)
...
reading u-boot.img

U-Boot 2013.04-dirty (Jul 10 2013 - 14:02:53)

I2C:   ready
DRAM:  512 MiB
WARNING: Caches not enabled
NAND:  No NAND device found!!!
0 MiB
MMC:   OMAP SD/MMC: 0, OMAP SD/MMC: 1
*** Warning - readenv() failed, using default environment
...
Hit any key to stop autoboot:  1 ... 0
gpio: pin 53 (gpio 53) value is 1
mmc0 is current device
micro SD card found
mmc0 is current device
gpio: pin 54 (gpio 54) value is 1
SD/MMC found on device 0
reading uEnv.txt
1672 bytes read in 7 ms (232.4 KiB/s)
Loaded environment from uEnv.txt
Importing environment from mmc ...
Running uenvcmd ...
reading zImage
```

```
4103240 bytes read in 470 ms (8.3 MiB/s)
reading initrd.img
2957458 bytes read in 341 ms (8.3 MiB/s)
reading /dtbs/am335x-boneblack.dtb
25926 bytes read in 10 ms (2.5 MiB/s)
## Flattened Device Tree blob at 88000000
   Booting using the fdt blob at 0x88000000
   Using Device Tree in place at 88000000, end 88009545
Starting kernel ...
Uncompressing Linux... done, booting the kernel.
...
Loading, please wait...
Scanning for Btrfs filesystems
systemd-fsck[216]: rootfs: clean, 87588/229216 files, 470285/928512 blocks

Debian GNU/Linux 7 beaglebone tty00

default username:password is [debian:temppwd]

Support/FAQ: http://elinux.org/Beagleboard:BeagleBoneBlack_Debian

The IP Address for usb0 is: 192.168.7.2
beaglebone login:
```

 You are now attached to the console serial port on the Bone, where system messages are logged. This is handy to have while doing low-level debugging. Watch out, though: those message could pop up any time and hide your login prompt.

5.6 Verifying You Have the Latest Version of the OS on Your Bone from the Shell

Problem

You are logged in to your Bone with a command prompt and want to know what version of the OS you are running.

Solution

Log in to your Bone and enter the following command:

```
bone# cat /etc/dogtag
BeagleBoard.org BeagleBone Debian Image 2014-11-11
```

Discussion

Recipe 1.3 shows how to open the *ID.txt* file to see the OS version. The */etc/dogtag* file has the same contents and is easier to find if you already have a command prompt. See Recipe 1.10 if you need to update your OS.

5.7 Controlling the Bone Remotely with VNC

Problem

You want to access the BeagleBone's graphical desktop from your host computer.

Solution

Run the installed Virtual Network Computing (VNC) server:

```
bone# sudo -u debian tightvncserver

You will require a password to access your desktops.

Password:
Verify:
Would you like to enter a view-only password (y/n)? n
xauth: (argv):1:  bad display name "beaglebone:1" in "add" command

New 'X' desktop is beaglebone:1

Creating default startup script /root/.vnc/xstartup
Starting applications specified in /root/.vnc/xstartup
Log file is /root/.vnc/beaglebone:1.log
```

 This solution assumes that you are logged in to the Bone as root. If you just started tightvncserver without sudo, the VNC would be running as root, which could cause some problems, because some programs (such as the Chrome web browser) won't run as root. This example uses sudo to run tightvncserver as debian.

To connect to the Bone, you will need to run a VNC client. There are many to choose from. Remmina Remote Desktop Client is already installed on Ubuntu. Start and select the new remote desktop file button (Figure 5-5).

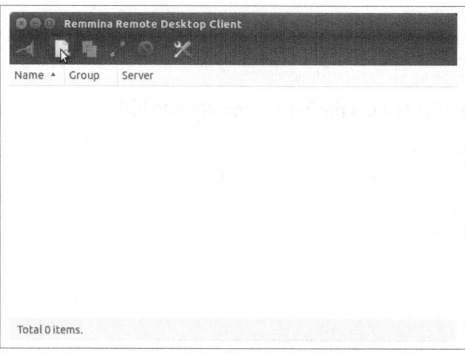

Figure 5-5. Creating a new remote desktop file in Remmina Remote Desktop Client

Give your connection a name, being sure to select "VNC - Virtual Network Computing." Also, be sure to add :1 after the server address, as shown in Figure 5-6. This should match the :1 that was displayed when you started vncserver.

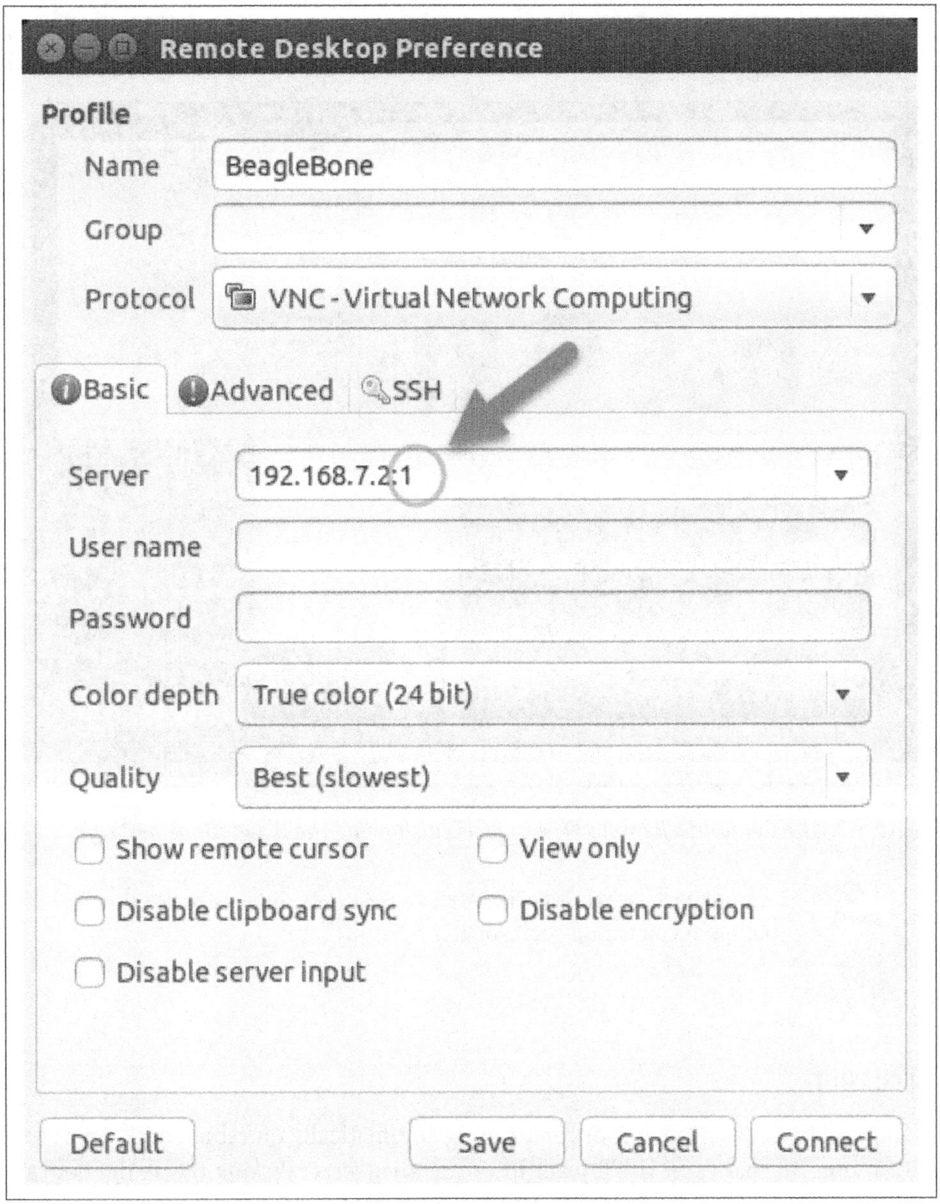

Figure 5-6. Configuring the Remmina Remote Desktop Client

Click Connect to start graphical access to your Bone, as shown in Figure 5-7.

Figure 5-7. The Remmina Remote Desktop Client showing the BeagleBone desktop

You might need to resize the VNC screen on your host to see the bottom menu bar on your Bone.

Discussion

Often, you'll want to have `vncserver` running automatically on reboot and run as the debian user. In that case, you'll want to create a *vncserver.desktop* file as the debian user and put it in the `autostart` folder for the window manager:

```
root@beaglebone:~# su - debian
debian@beaglebone:~$ mkdir -p .config/autostart
debian@beaglebone:~$ cat > .config/autostart/vncserver.desktop
[Desktop Entry]
Type=Application
Name=vncserver
Exec=vncserver :1
```

```
StartupNotify=false
^D
debian@beaglebone:~$ vncpasswd
Using password file /home/debian/.vnc/passwd
Password: password
Verify:   password
Would you like to enter a view-only password (y/n)? n
debian@beaglebone:~$ logout
root@beaglebone:~# /etc/init.d/lightdm restart
Stopping Light Display Manager: lightdm.
Starting Light Display Manager: lightdm.
root@beaglebone:~#
```

5.8 Learning Typical GNU/Linux Commands

Problem

There are many powerful commands to use in Linux. How do you learn about them?

Solution

Table 5-1 lists many common Linux commands.

Table 5-1. Common Linux commands

Command	Action
pwd	show current directory
cd	change current directory
ls	list directory contents
chmod	change file permissions
chown	change file ownership
cp	copy files
mv	move files
rm	remove files
mkdir	make directory
rmdir	remove directory

Command	Action
cat	dump file contents
less	progressively dump file
vi	edit file (complex)
nano	edit file (simple)
head	trim dump to top
tail	trim dump to bottom
echo	print/dump value
env	dump environment variables
export	set environment variable
history	dump command history
grep	search dump for strings
man	get help on command
apropos	show list of man pages
find	search for files
tar	create/extract file archives
gzip	compress a file
gunzip	decompress a file
du	show disk usage
df	show disk free space
mount	mount disks
tee	write dump to file in parallel
hexdump	readable binary dumps
whereis	locates binary and source files

Discussion

You'll find many good resources online to learn Linux commands. A good place to begin is Free Electrons, which has a nice one-page summary of many useful commands (*http://bit.ly/1GEQglM*). It also has a nice set of slides that provide details of the commands (*http://bit.ly/1E5wAkT*).

5.9 Editing a Text File from the GNU/Linux Command Shell

Problem

You want to run an editor to change a file.

Solution

The Bone comes with a number of editors. The simplest to learn is nano. Just enter the following command:

```
bone# nano file
```

You are now in nano (Figure 5-8). You can't move around the screen using the mouse, so use the arrow keys. The bottom two lines of the screen list some useful commands. Pressing ∧G (Ctrl-G) will display more useful commands. ∧X (Ctrl-X) exits nano and gives you the option of saving the file.

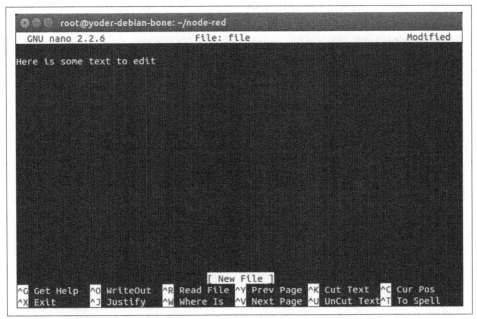

Figure 5-8. Editing a file with nano

By default, the file you create will be saved in the directory from which you opened nano.

Discussion

Many other text editors will run on the Bone. vi, vim, emacs, and even eclipse are all supported. See Recipe 5.15 to learn if your favorite is one of them.

5.10 Using a Graphical Editor

Problem

You want to use a graphical (rather than text-based) editor to edit a text file.

Solution

leafpad is a simple graphics-based editor that's already installed on your Bone. You can use it if your host computer supports X-Windows (Mac and Linux). Just run the following commands to open a new file to begin editing in leafpad (Figure 5-9):

```
host$ ssh -X root@192.168.7.2
bone# leafpad file.txt
```

Notice the -X on the ssh command. This instructs SSH to pass the graphics from the Bone to the host computer.

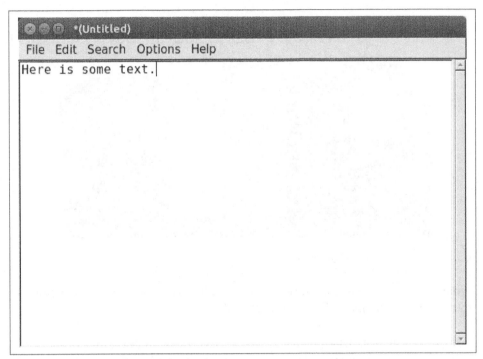

Figure 5-9. Editing a file with leafpad

Now, you can use your mouse to move around and copy and paste as usual.

5.11 Establishing an Ethernet-Based Internet Connection

Problem

You want to connect your Bone to the Internet using the wired network connection.

Solution

Plug one end of an Ethernet patch cable into the RJ45 connector on the Bone (see Figure 5-10) and the other end into your home hub/router. The yellow and green link lights on both ends should begin to flash.

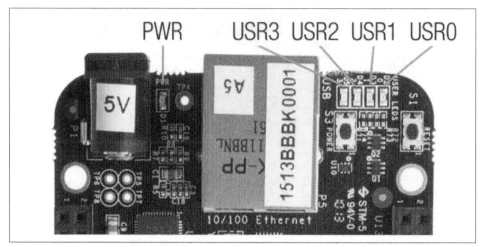

Figure 5-10. The RJ45 port on the Bone

If your router is already configured to run DHCP (Dynamical Host Configuration Protocol), it will automatically assign an IP address to the Bone.

 It might take a minute or two for your router to detect the Bone and assign the IP address.

To find the IP address, open a terminal window and run the ifconfig command:

```
bone# ifconfig
eth0      Link encap:Ethernet  HWaddr c8:a0:30:b7:f9:6f
          inet addr:137.112.41.36  Bcast:137.112.41.255  Mask:255.255.255.0
          inet6 addr: fe80::caa0:30ff:feb7:f96f/64 Scope:Link
          UP BROADCAST RUNNING MULTICAST  MTU:1500  Metric:1
          RX packets:238 errors:0 dropped:53 overruns:0 frame:0
          TX packets:20 errors:0 dropped:0 overruns:0 carrier:0
          collisions:0 txqueuelen:1000
          RX bytes:25341 (24.7 KiB)  TX bytes:4339 (4.2 KiB)
          Interrupt:40

lo        Link encap:Local Loopback
          inet addr:127.0.0.1  Mask:255.0.0.0
          inet6 addr: ::1/128 Scope:Host
          UP LOOPBACK RUNNING  MTU:65536  Metric:1
          RX packets:1517 errors:0 dropped:0 overruns:0 frame:0
          TX packets:1517 errors:0 dropped:0 overruns:0 carrier:0
          collisions:0 txqueuelen:0
          RX bytes:2542320 (2.4 MiB)  TX bytes:2542320 (2.4 MiB)
```

```
usb0      Link encap:Ethernet  HWaddr aa:26:91:0e:a6:17
          inet addr:192.168.7.2  Bcast:192.168.7.3  Mask:255.255.255.252
          inet6 addr: fe80::a826:91ff:fe0e:a617/64 Scope:Link
          UP BROADCAST RUNNING MULTICAST  MTU:1500  Metric:1
          RX packets:8672 errors:0 dropped:0 overruns:0 frame:0
          TX packets:5069 errors:0 dropped:0 overruns:0 carrier:0
          collisions:0 txqueuelen:1000
          RX bytes:7474185 (7.1 MiB)  TX bytes:3262782 (3.1 MiB)
```

My Bone is connected to the Internet in two ways: via the RJ45 connection (eth0) and via the USB cable (usb0). The inet addr field shows that my Internet address is 137.112.41.36 for the RJ45 connector.

On my university campus, you must register your MAC address before any device will work on the network. The HWaddr field gives the MAC address. For eth0, it's c8:a0:30:b7:f9:6f.

The IP address of your Bone can change. If it's been assigned by DHCP, it can change at any time. The MAC address, however, never changes; it is assigned to your ethernet device when it's manufactured.

When a Bone is connected to some networks (in my case, the campus network), it becomes visible to the *world*. If you don't secure your Bone, the world will soon find it. See "Changing passwords" on page 153 and Recipe 5.14.

On many home networks, you will be behind a firewall and won't be as visible.

Discussion

At this point, it's a good idea to change your passwords and your Bone's hostname.

Changing passwords

Be sure to change *both* of your Bone's passwords (root and debian). Log in as root and run passwd:

```
bone# passwd
Enter new UNIX password:
Retype new UNIX password:
passwd: password updated successfully
```

Then log in at debian and run passwd again.

Setting your Bone's hostname

If you are on a network with other BeagleBones, it's a good idea to give your Bone a unique name. The default name is beaglebone. Assigning a new name is as easy as running the following command:

```
bone# echo "newname" > /etc/hostname
```

Also, edit /etc/hosts and change all occurrences of *beaglebone* to the new network name. This changes the network name to newname. When picking a name, don't use punctuation or other special characters. Even _ isn't allowed. The new name will appear after you reboot.

5.12 Establishing a WiFi-Based Internet Connection

Problem

You want BeagleBone Black to talk to the Internet using a USB wireless adapter.

Solution

Several WiFi adapters work with the Bone. Check WiFi Adapters (*http://bit.ly/ 1EbEwUo*) for the latest list.

To make this recipe, you will need:

- USB Wifi adapter (see "Miscellaneous" on page 318)
- 5 V external power supply (see "Miscellaneous" on page 318)

 Most adapters need at least 1 A of current to run, and USB supplies only 0.5 A, so be sure to use an external power supply. Otherwise, you will experience erratic behavior and random crashes.

First, plug in the WiFi adapter and the 5 V external power supply and reboot.

Then run lsusb to ensure that your Bone found the adapter:

```
bone# lsusb
Bus 001 Device 002: ID 0bda:8176 Realtek Semiconductor Corp. RTL8188CUS 802.11n
WLAN Adapter
Bus 001 Device 001: ID 1d6b:0002 Linux Foundation 2.0 root hub
Bus 002 Device 001: ID 1d6b:0002 Linux Foundation 2.0 root hub
```

There is a well-known bug in the Bone's 3.8 kernel series that prevents USB devices from being discovered when hot-plugged, which is why you should reboot. Newer kernels should address this issue.

Next, run `iwconfig` to find your adapter's name. Mine is called `wlan0`, but you might see other other names, such as `ra0`.

```
bone# iwconfig
wlan0     unassociated  Nickname:"WIFI@REALTEK"
          Mode:Auto  Frequency=2.412 GHz  Access Point: Not-Associated
          Sensitivity:0/0
          Retry:off   RTS thr:off   Fragment thr:off
          Encryption key:off
          Power Management:off
          Link Quality=0/100  Signal level=0 dBm  Noise level=0 dBm
          Rx invalid nwid:0  Rx invalid crypt:0  Rx invalid frag:0
          Tx excessive retries:0  Invalid misc:0   Missed beacon:0

lo        no wireless extensions.

eth0      no wireless extensions.

usb0      no wireless extensions.
```

If no name appears, try `ifconfig -a`:

```
bone# ifconfig -a
...
usb0      Link encap:Ethernet  HWaddr 92:84:3f:a2:12:0f
          inet addr:192.168.7.2  Bcast:192.168.7.3  Mask:255.255.255.252
          inet6 addr: fe80::9084:3fff:fea2:120f/64 Scope:Link
          UP BROADCAST RUNNING MULTICAST  MTU:1500  Metric:1
          RX packets:5789 errors:0 dropped:0 overruns:0 frame:0
          TX packets:5158 errors:0 dropped:0 overruns:0 carrier:0
          collisions:0 txqueuelen:1000
          RX bytes:6401357 (6.1 MiB)  TX bytes:842982 (823.2 KiB)

wlan0     Link encap:Ethernet  HWaddr 00:13:ef:d0:27:04
          UP BROADCAST MULTICAST  MTU:1500  Metric:1
          RX packets:0 errors:0 dropped:0 overruns:0 frame:0
          TX packets:0 errors:0 dropped:0 overruns:0 carrier:0
          collisions:0 txqueuelen:1000
          RX bytes:0 (0.0 B)  TX bytes:0 (0.0 B)
```

When you find the WiFi device name (mine is `wlan0`), remember it for later.

Run `wicd-curses` to display the wireless networks your WiFi device is seeing:

```
bone# wicd-curses
```

In Figure 5-11, you see one wireless network, my home wireless.

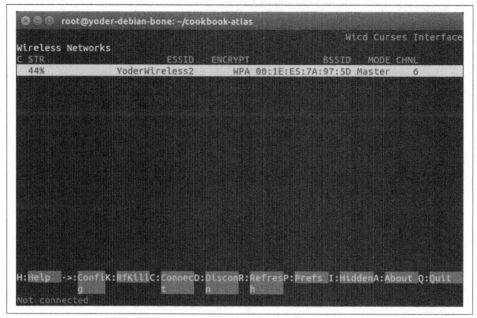

Figure 5-11. Using wicd to display available wireless networks

If nothing appears, press P and make sure the Wireless Interface matches the device name you noted from `ifconfig`. If it doesn't match, press the down arrow and enter the device name, as shown in Figure 5-12.

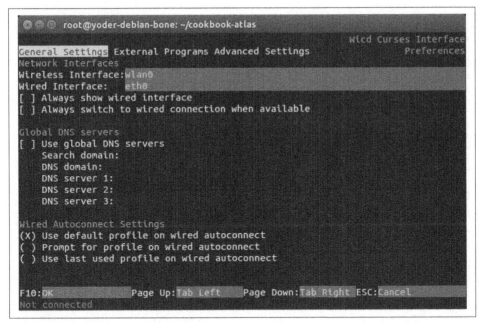

Figure 5-12. Setting preferences in wicd

After you enter a name, press F10 to save and then press R to refresh the list. You'll now see the available access points.

Use the down arrow to select which network to use and then press the right arrow to configure it, as shown in Figure 5-13.

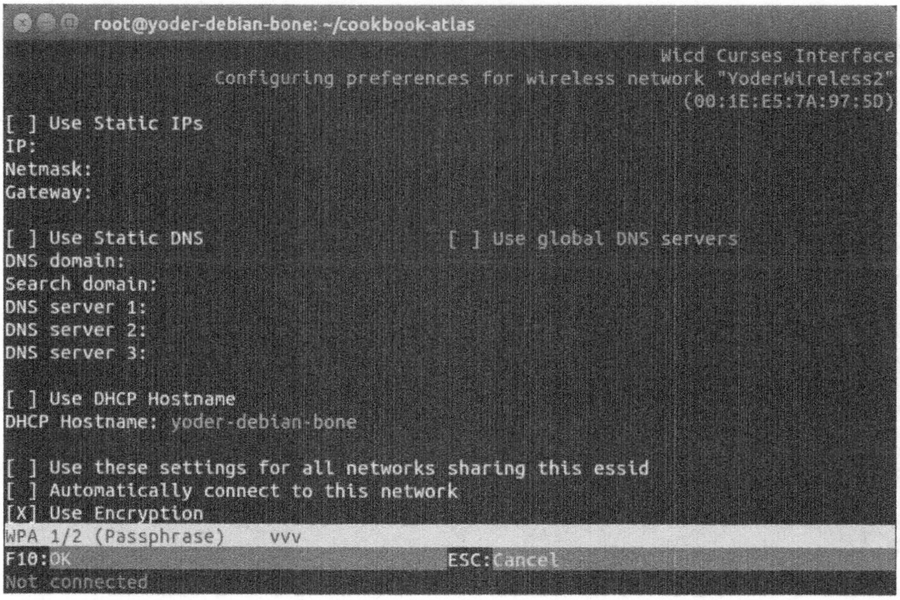

Figure 5-13. Configuring a network in wicd

Press the down arrow until you are near the bottom of the screen and then press Enter. Use the up/down arrows to select the type of authentication (as shown in Figure 5-14) and then press Enter. My home system uses WPA 1/2 (Passphrase) (Figure 5-15).

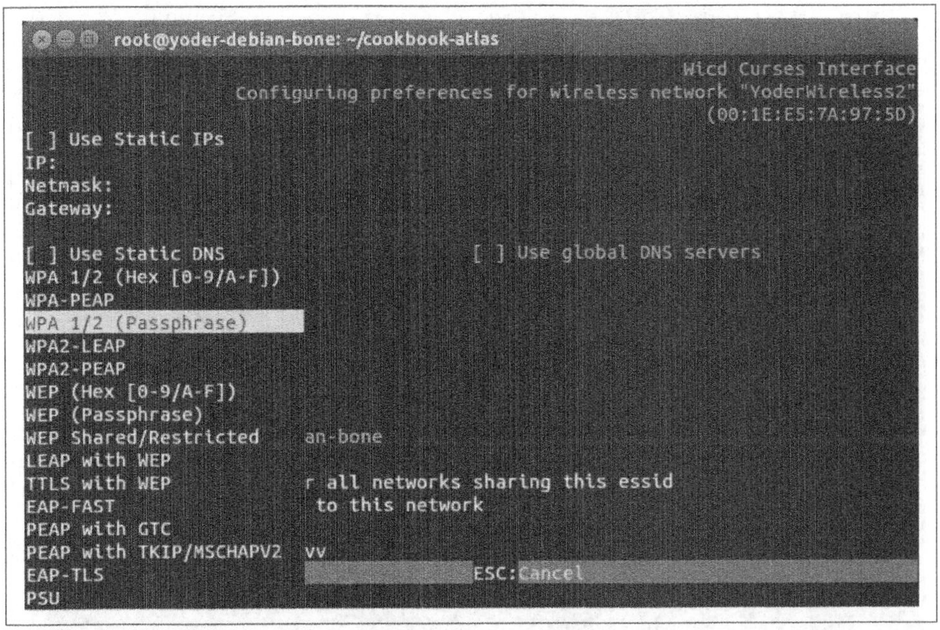

Figure 5-14. Configuring authentication in wicd

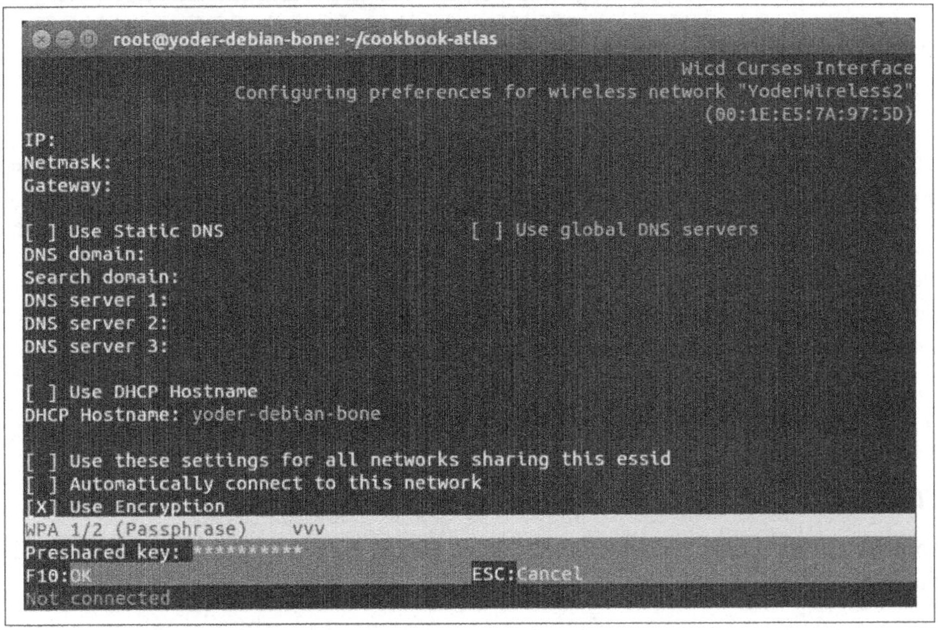

Figure 5-15. Entering password in wicd

Press F10 and select the network again; then press C to connect. You'll be connected in a moment, as shown in Figure 5-16.

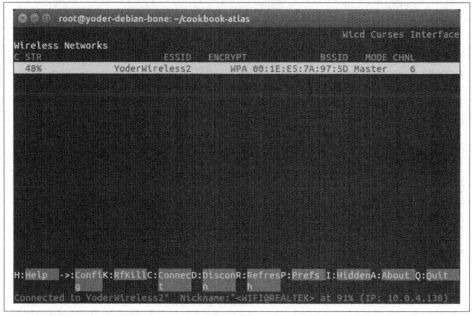

Figure 5-16. Connected to a network via wicd

Discussion

Congratulations, you are running wirelessly. Your Bone's IP address is shown in the bottom of the window (mine is 10.0.4.138). You can press Q to quit, and your connection will continue.

Now that you are connected wirelessly, you can use the Bone to gather remote data. Just wire your sensors to the Bone, and it can send the sensor data to anywhere on the Internet via the wireless connection.

5.13 Sharing the Host's Internet Connection over USB

Problem

Your host computer is connected to the Bone via the USB cable, and you want to run the network between the two.

Solution

Recipe 5.11 shows how to connect BeagleBone Black to the Internet via the RJ45 Ethernet connector. This recipe shows a way to connect without using the RJ45 connector.

A network is automatically running between the Bone and the host computer at boot time using the USB. The host's IP address is 192.168.7.1 and the Bone's is 192.168.7.2. Although your Bone is talking to your host, it can't reach the Internet in general, nor can the Internet reach it. On one hand, this is good, because those who are up to no good can't access your Bone. On the other hand, your Bone can't reach the rest of the world.

Letting your bone see the world: setting up IP masquerading

You need to set up IP masquerading on your host and configure your Bone to use it. Here is a solution that works with a host computer running Linux. Add the code in Example 5-1 to a file called *ipMasquerade.sh* on your host computer.

Example 5-1. Code for IP Masquerading (ipMasquerade.sh)

```
#!/bin/bash
# These are the commands to run on the host to set up IP
#  masquerading so the Bone can access the Internet through
#  the USB connection.
# This configures the host, run ./setDNS.sh to configure the Bone.
# Inspired by http://thoughtshubham.blogspot.com/2010/03/
#  internet-over-usb-otg-on-beagleboard.html

if [ $# -eq 0 ] ; then
echo "Usage: $0 interface (such as eth0 or wlan0)"
exit 1
fi

interface=$1
hostAddr=192.168.7.1
beagleAddr=192.168.7.2
ip_forward=/proc/sys/net/ipv4/ip_forward

if [ `cat $ip_forward` == 0 ]
  then
    echo "You need to set IP forwarding. Edit /etc/sysctl.conf using:"
    echo "$ sudo nano /etc/sysctl.conf"
    echo "and uncomment the line    \"net.ipv4.ip_forward=1\""
    echo "to enable forwarding of packets. Then run the following:"
    echo "$ sudo sysctl -p"
    exit 1
  else
    echo "IP forwarding is set on host."
```

```
fi
# Set up IP masquerading on the host so the bone can reach the outside world
sudo iptables -t nat -A POSTROUTING -s $beagleAddr -o $interface -j MASQUERADE
```

Then, on your host, run the following commands:

```
host$ chmod +x ipMasquerade.sh
host$ ./ipMasquerade.sh eth0
```

This will direct your host to take requests from the Bone and send them to eth0. If your host is using a wireless connection, change eth0 to wlan0.

Now let's set up your host to instruct the Bone what to do. Add the code in Example 5-2 to *setDNS.sh* on your host computer.

Example 5-2. Code for setting the DNS on the Bone (setDNS.sh)

```
#!/bin/bash
# These are the commands to run on the host so the Bone
#  can access the Internet through the USB connection.
# Run ./ipMasquerade.sh the first time. It will set up the host.
# Run this script if the host is already set up.

hostAddr=192.168.7.1
beagleAddr=192.168.7.2

# Save the /etc/resolv.conf on the Bone in case we mess things up.
ssh root@$beagleAddr "mv -n /etc/resolv.conf /etc/resolv.conf.orig"
# Create our own resolv.conf
cat - << EOF > /tmp/resolv.conf
# This is installed by host.setDNS.sh on the host

EOF

# Look up the nameserver of the host and add it to our resolv.conf
nmcli dev list | grep IP4.DNS | sed 's/IP4.DNS\[.\]:/nameserver/' >> /tmp/resolv.conf
scp /tmp/resolv.conf root@$beagleAddr:/etc

# Tell the beagle to use the host as the gateway.
ssh root@$beagleAddr "/sbin/route add default gw $hostAddr" || true
```

Then, on your host, run the following commands:

```
host$ chmod +x setDNS.sh
host$ ./setDNS.sh
host$ ssh -X root@192.168.7.2
bone# ping -c2 google.com
PING google.com (216.58.216.96) 56(84) bytes of data.
64 bytes from ord30s22....net (216.58.216.96): icmp_req=1 ttl=55 time=7.49 ms
64 bytes from ord30s22....net (216.58.216.96): icmp_req=2 ttl=55 time=7.62 ms

--- google.com ping statistics ---
```

```
2 packets transmitted, 2 received, 0% packet loss, time 1002ms
rtt min/avg/max/mdev = 7.496/7.559/7.623/0.107 ms
```

This will look up what Domain Name System (DNS) servers your host is using and copy them to the right place on the Bone. The ping command is a quick way to verify your connection.

Letting the world see your bone: setting up port forwarding

Now your Bone can access the world via the USB port and your host computer, but what if you have a web server on your Bone that you want to access from the world? The solution is to use *port forwarding* from your host. Web servers typically listen to port 80. First, look up the IP address of your host:

```
host$ ifconfig
eth0      Link encap:Ethernet  HWaddr 00:e0:4e:00:22:51
          inet addr:137.112.41.35  Bcast:137.112.41.255  Mask:255.255.255.0
          inet6 addr: fe80::2e0:4eff:fe00:2251/64 Scope:Link
          UP BROADCAST RUNNING MULTICAST  MTU:1500  Metric:1
          RX packets:5371019 errors:0 dropped:0 overruns:0 frame:0
          TX packets:4720856 errors:0 dropped:0 overruns:0 carrier:0
          collisions:0 txqueuelen:1000
          RX bytes:1667916614 (1.6 GB)  TX bytes:597909671 (597.9 MB)

eth1      Link encap:Ethernet  HWaddr 00:1d:60:40:58:e6
...
```

It's the number following inet addr:, which in my case is 137.112.41.35.

> If you are on a wireless network, find the IP address associated with wlan0.

Then run the following, using your host's IP address:

```
host$ sudo iptables -t nat -A PREROUTING -p tcp -s 0/0 \
      -d 137.112.41.35 --dport 1080 -j DNAT --to 192.168.7.2:80
```

Now browse to your host computer at port 1080. That is, if your host's IP address is 123.456.789.0, enter 123.456.789.0:1080. The :1080 specifies what port number to use. The request will be forwarded to the server on your Bone listening to port 80. (I used 1080 here, in case your host is running a web server of its own on port 80.)

Discussion

Whenever a device connects to another device on the Internet, two numbers are needed: the *IP address* (192.168.7.2 for example) and the *port* number. If you are

---

using SSH to connect, it defaults to port 22; if it's a web browser, it uses port 80 (or 443 if it's encrypted); and so on. You can even write your own program to listen on a given port. The `iptables` command given here says to take any requests to connect to port 1080 (`--dport 1080`) on 137.112.41.35 and forward them to 192.168.7.2, port 80. It's now as if the request were made to connect to 192.168.7.2 instead.

 Be careful, you've now exposed your Bone to the world. Anyone from the outside can access your Bone. See Recipe 5.14 for setting up a firewall to limit which IP addresses can access your Bone.

If you change your mind about the port forwarding, switching the `-A` to `-D` will delete the entry and stop the forwarding. Rebooting will also stop the forwarding.

```
host$ sudo iptables -t nat -D PREROUTING -p tcp -s 0/0 \
      -d 137.112.41.35 --dport 1080 -j DNAT --to 192.168.7.2:80
```

5.14 Setting Up a Firewall

Problem

You have put your Bone on the network and want to limit which IP addresses can access it.

Solution

The Bone already has a firewall installed. Just run the following command:

```
bone# iptables --policy INPUT DROP
bone# iptables -A INPUT -s 137.112.0.0/16 -j ACCEPT
```

In this case, only users from my campus (IP addresses beginning with 137.112) will be able to access (or even see) your Bone.

Discussion

The first `iptables` command says to drop all access from the outside world. The second line then lists an exception, allowing any IP address beginning with 137.112 to have access. The /16 says to let the last 16 bits of the IP have any value. If I wanted to limit access to only those on my subnet, I could use 137.112.41.0/8. Now, only the last 8 bits can be anything.

You can remove an exception by using the following command:

```
bone# iptables -D INPUT -s 137.112.0.0/16 -j ACCEPT
```

Or just reboot. You have to rerun `iptables` every time you boot your Bone.

5.15 Installing Additional Packages from the Debian Package Feed

Problem

You want to do more cool things with your BeagleBone by installing more programs.

Solution

Your Bone needs to be on the network for this to work. See Recipe 5.11, Recipe 5.12, or Recipe 5.13.

The easiest way to install more software is to use `apt-get`:

```
bone# apt-get update
bone# apt-get install "name of software"
```

A `sudo` is necessary if you aren't running as `root`. The first command downloads package lists from various repositories and updates them to get information on the newest versions of packages and their dependencies. (You need to run it only once a week or so.) The second command fetches the software and installs it and all packages it depends on.

How do you find out what software you can install? Try running this:

```
bone# apt-cache pkgnames | sort > /tmp/list
bone# wc /tmp/list
 32121   32121 502707 /tmp/list
bone# less /tmp/list
```

The first command lists all the packages that `apt-get` knows about and sorts them and stores them in */tmp/list*. The second command shows why you want to put the list in a file. The `wc` command counts the number of lines, words, and characters in a file. In our case, there are over 32,000 packages from which we can choose! The `less` command displays the sorted list, one page at a time. Press the space bar to go to the next page. Press Q to quit.

Suppose that you would like to install an online dictionary (`dict`). Just run the following command:

```
bone# apt-get install dict
```

Now you can run `dict`.

Discussion

The Debian package manager has a large number of pretested packages that you can load and run on the Bone. The Debian packages page (*http://bit.ly/1KWoR1A*) provides details about the packages in the current release of Debian. For example, the package page for git (*http://bit.ly/1HC85zm*) describes git and identifies the version that will be loaded by running apt-get install git. The version that apt-get installs might not be the most current version, but it is well tested.

What if you would like to run a newer version of a package? Debian has different releases of packages that reflect different levels of testing. When a package is first entered into Debian, it is put in the Unstable release category. It lives here for a few days until it is tested on the currently supported architectures and appears not to have any obvious bugs. It is then moved to a Testing release, where further testing is done and bugs can be found. Finally, a new package can move to the Stable release if it is a significant enough upgrade.

At the time of this writing, Wheezy (version 7) is the current stable release, and Jessie (version 8) is the Testing release. You can install packages from Jessie, but watch out: there is a reason why it's called Testing.

For example, Wheezy uses version 1.7 of git (*http://bit.ly/1HC85zm*), but version 1.8 of git introduced some features that you want to use. Jessie uses version 2.0 of git (*http://bit.ly/1KWprfS*), so you could upgrade to it (details in a moment).

What if you don't want to risk upgrading a Testing release? You could use Wheezy-backports. Backports (*http://backports.debian.org/*) are packages taken from the next Debian release, adjusted and recompiled for use on the current release. For this example, version 1.9 of git (*http://bit.ly/1B4uGyW*) is used in Wheezy-backports.

To load either the Wheezy-backports or the Jessie packages, first run the following command to direct apt-get where to get the new packages:

```
bone# cd /etc/apt/sources.list.d
bone# echo "deb ftp://ftp.debian.org/debian/ wheezy-backports main
deb-src ftp://ftp.debian.org/debian/ wheezy-backports main" > \
wheezy-backports.list
bone# echo "deb ftp://ftp.debian.org/debian/ jessie main
deb-src ftp://ftp.debian.org/debian/ jessie main" > jessie.list
bone# echo 'APT::Default-Release "stable";' > /etc/apt/apt.conf.d/local

bone# apt-get update
```

The second-to-last command makes Wheezy the default release to use. The last command will go out to the repositories and load information on all the new packages. It might take awhile.

Be sure to run the apt-get update; otherwise, newly added repositories won't be seen.

Then, to install the Wheezy-backports version, run the following command:

```
bone# apt-get install git/wheezy-backports
bone# git --version
git version 1.9.1
```

Or, to install the Jessie version, run the following command:

```
bone# apt-get install git/jessie
git version 2.1.1
```

Note that you might need to install other packages first. If you get an error, see what
is missing and try installing it before trying again.

Adding all the other releases will make apt-get run slower. Here is
how I like to disable them:

```
bone# cd /etc/apt/sources.list.d
bone# ls
jessie.list  sid.list  wheezy-backports.list
bone# mkdir hide
bone# mv *.list hide
```

This moves the files you created earlier into a directory called *hide*,
where apt-get won't see them. apt-get will now run as before.
Later, if needed, you can move them out of the *hide* directory.

```
bone# cd /etc/apt/sources.list.d
bone# mv hide/jessie.list .
```

5.16 Removing Packages Installed with apt-get

Problem

You've been playing around and installing all sorts of things with apt-get and now
you want to clean things up a bit.

Solution

apt-get has a remove option, so you can run the following command:

```
bone# apt-get remove dict
Reading package lists... Done
Building dependency tree
Reading state information... Done
```

```
The following packages were automatically installed and are no longer required:
  libmaa3 librecode0 recode
Use 'apt-get autoremove' to remove them.
The following packages will be REMOVED:
  dict
0 upgraded, 0 newly installed, 1 to remove and 27 not upgraded.
After this operation, 164 kB disk space will be freed.
Do you want to continue [Y/n]? y
```

Discussion

Removing packages with remove doesn't remove everything, because prerequisite packages might have also been installed. You can remove everything with autoremove:

```
bone# apt-get autoremove dict
Reading package lists... Done
Building dependency tree
Reading state information... Done
Package 'dict' is not installed, so not removed
The following packages will be REMOVED:
  libmaa3 librecode0 recode
0 upgraded, 0 newly installed, 3 to remove and 27 not upgraded.
After this operation, 2,201 kB disk space will be freed.
Do you want to continue [Y/n]? y
```

Notice an additional 2,201 kB (more than dict itself) was removed.

5.17 Copying Files Between the Onboard Flash and the MicroSD Card

Problem

You want to move files between the onboard flash and the microSD card.

Solution

If you booted from the microSD card, run the following command:

```
bone# df -h
Filesystem       Size  Used Avail Use% Mounted on
rootfs           7.2G  2.0G  4.9G  29% /
udev              10M     0   10M   0% /dev
tmpfs            100M  1.9M   98M   2% /run
/dev/mmcblk0p2   7.2G  2.0G  4.9G  29% /
tmpfs            249M     0  249M   0% /dev/shm
tmpfs            249M     0  249M   0% /sys/fs/cgroup
tmpfs            5.0M     0  5.0M   0% /run/lock
tmpfs            100M     0  100M   0% /run/user
bone# ls /dev/mmcblk*
```

```
/dev/mmcblk0      /dev/mmcblk0p2   /dev/mmcblk1boot0   /dev/mmcblk1p1
/dev/mmcblk0p1    /dev/mmcblk1     /dev/mmcblk1boot1
```

The df command shows what partitions are already mounted. The line /dev/
mmcblk0p2 7.2G 2.0G 4.9G 29% / shows that mmcblk0 partition p2 is mounted as /,
the root file system. The general rule is that the media you're booted from (either the
onboard flash or the microSD card) will appear as mmcblk0. The second partition (p2)
is the root of the file system.

The ls command shows what devices are available to mount. Because mmcblk0 is
already mounted, /dev/mmcblk1p1 must be the other media that we need to mount.
Run the following commands to mount it:

```
bone# cd /mnt
bone# mkdir onboard
bone# ls onboard
bone# mount /dev/mmcblk1p1 onboard/
bone# ls onboard
bin    etc     lib         mnt            proc  sbin     sys  var
boot   home    lost+found  nfs-uEnv.txt   root  selinux  tmp
dev    ID.txt  media       opt            run   srv      usr
```

The cd command takes us to a place in the file system where files are commonly
mounted. The mkdir command creates a new directory (*onboard*) to be a mount
point. The ls command shows there is nothing in *onboard*. The mount command
makes the contents of the onboard flash accessible. The next ls shows there now are
files in *onboard*. These are the contents of the onboard flash, which can be copied to
and from like any other file.

Discussion

This same process should also work if you have booted from the onboard flash.
When you are done with the onboard flash, you can unmount it by using this command:

```
bone# umount /mnt/onboard
```

5.18 Freeing Space on the Onboard Flash or MicroSD Card

Problem

You are starting to run out of room on your microSD card (or onboard flash) and
have removed several packages you had previously installed (Recipe 5.16), but you
still need to free up more space.

Solution

To free up space, you can remove preinstalled packages or discover big files to remove.

Removing preinstalled packages

You might not need a few things that come preinstalled in the Debian image, including such things as OpenCV, the Chromium web browser, and some documentation.

The Chromium web browser is the open source version of Google's Chrome web browser. Unless you are using the Bone as a desktop computer, you can probably remove it.

OpencCV (*http://opencv.org/*) is an open source computer vision package. If you aren't doing computer vision, you can remove it.

Here's how you can remove these:

```
bone# apt-get autoremove opencv*        (83M)
bone# apt-get remove libopencv-* --purge  (41M)
bone# apt-get autoremove                (42M)
bone# rm -rf /usr/lib/chromium          (95M)
bone# rm -f  /usr/bin/chromium
bone# rm -rf /usr/share/doc             (101M)
bone# rm -rf /usr/share/man             (27M)
```

These packages and sizes to remove were found in a BeagleBoard Google Groups thread (*http://bit.ly/1HC96aI*).

Discovering big files

The du (disk usage) command offers a quick way to discover big files:

```
bone# du -shx /*
5.6M    /bin
11M     /boot
0       /dev
5.4M    /etc
232K    /home
66M     /lib
16K     /lost+found
388M    /media
4.0K    /mnt
63M     /opt
du: cannot access `/proc/20551/task/20551/fd/3': No such file or directory
du: cannot access `/proc/20551/task/20551/fdinfo/3': No such file or directory
```

```
du: cannot access `/proc/20551/fd/3': No such file or directory
du: cannot access `/proc/20551/fdinfo/3': No such file or directory
0         /proc
227M      /root
1.1M      /run
5.0M      /sbin
4.0K      /selinux
4.0K      /srv
0         /sys
36K       /tmp
1.2G      /usr
143M      /var
```

If you booted from the microSD card, du lists the usage of the microSD. If you booted from the onboard flash, it lists the onboard flash usage.

The -s option summarizes the results rather than displaying every file. -h prints it in *human* form—that is, using M and K postfixes rather than showing lots of digits. The /* specifies to run it on everything in the top-level directory. It looks like a couple of things disappeared while the command was running and thus produced some error messages.

 For more help, try du --help.

The */usr* directory appears to be the biggest user of space at 1.2 GB. You can then run the following command to see what's taking up the space in */usr*:

```
bone# du -sh /usr/*
46M       /usr/bin
4.0K      /usr/games
36M       /usr/include
424M      /usr/lib
16M       /usr/local
4.0M      /usr/sbin
694M      /usr/share
4.0K      /usr/src
```

A more interactive way to explore your disk usage is by installing ncdu (ncurses disk usage):

```
bone# apt-get install ncdu
bone# ncdu /
```

After a moment, you'll see the following:

```
ncdu 1.8 ~ Use the arrow keys to navigate, press ? for help
--- / --------------------------------------------------------------
```

```
       1.2GiB [##########] /usr
     387.6MiB [###       ] /media
     226.0MiB [#         ] /root
     142.9MiB [#         ] /var
      65.9MiB [          ] /lib
      62.4MiB [          ] /opt
      10.8MiB [          ] /boot
       5.6MiB [          ] /bin
       5.4MiB [          ] /etc
       5.0MiB [          ] /sbin
       1.0MiB [          ] /run
     232.0KiB [          ] /home
      36.0KiB [          ] /tmp
   e  16.0KiB [          ] /lost+found
   e   4.0KiB [          ] /srv
   e   4.0KiB [          ] /selinux
   e   4.0KiB [          ] /mnt
       0.0  B [          ] /sys
       0.0  B [          ] /proc
       0.0  B [          ] /dev

 Total disk usage:   2.1GiB  Apparent size:   1.9GiB  Items: 156583
```

ncdu is a character-based graphics interface to du. You can now use your arrow keys to navigate the file structure to discover where the big unused files are. Press ? for help.

Be careful not to press the D key, because it's used to delete a file or directory.

Discussion

You can also list all the packages and their sizes by using the following command:

```
bone# dpkg-query -Wf '${Installed-Size}\t${Package}\n' | sort -rn | less
```

The less command is a handy way to view an output one screen at at time. Press the space bar to advance to the next page. Press D for half a page and Q to quit. Run less --help for more tips.

Just be careful what you remove. You might be using something that you think is safe to delete.

5.19 Installing Additional Node.js Packages

Problem

You are working with JavaScript and want to build on what others have done.

Solution

Node Package Manager (npm [*https://www.npmjs.org/*]) is used to install new packages. For example, suppose that you are wiring up a GPS (Recipe 2.8) and want to use the nmea JavaScript package to decode the National Marine Electronics Association (NMEA) data from the GPS. Install the nmea package with the following command:

```
bone# npm install -g nmea
npm http GET https://registry.npmjs.org/nmea
npm http 200 https://registry.npmjs.org/nmea
npm http GET https://registry.npmjs.org/nmea/-/nmea-0.0.7.tgz
npm http 200 https://registry.npmjs.org/nmea/-/nmea-0.0.7.tgz
nmea@0.0.7 /usr/local/lib/node_modules/nmea
```

The -g instructs to install globally. That way, you can use nmea from any directory. If -g is not given, you can use nmea only in the directory in which you installed it.

Discussion

Node.js (*http://nodejs.org/*) is a platform built on Chrome's JavaScript runtime (*http://bit.ly/1wFB3fE*). It's what runs the JavaScript we have been using. More than 112,000 packages have been written for Node.js and are available using npm. Visit the npm site (*https://www.npmjs.org/*) (Figure 5-17) to see what's available.

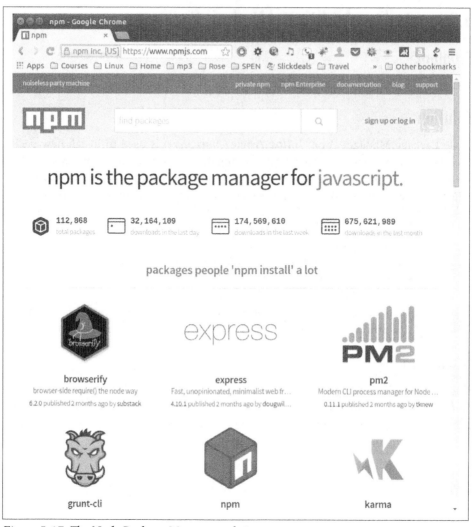

Figure 5-17. The Node Package Manager website

Use the web browser on your host computer to visit *npmjs.org*. When you find packages you want, go back to your Bone and use the npm install command to install them on the Bone.

To see what Node.js packages are installed, run the following command:

```
bone# npm list -g | less
/usr/local/lib
└─ bonescript@0.2.4
```

```
|  ├─ epoll@0.0.7
|  |  └─ nan@0.4.4
|  ├─ express@3.1.0
|  |  ├─ buffer-crc32@0.1.1
|  |  ├─ commander@0.6.1
|  |  ├─ connect@2.7.2
|  |  |  ├─ bytes@0.1.0
|  |  |  ├─ formidable@1.0.11
|  |  |  ├─ pause@0.0.1
|  |  |  └─ qs@0.5.1
...
```

The output provides a listing of what packages are installed (such as BoneScript) and what packages they depend on. Here, `bonescript` depends on `epoll`, which depends on `nan`, and so on.

5.20 Using Python to Interact with the Physical World

Problem

You want to use Python on the Bone to talk to the world.

Solution

So far, we've focused on using JavaScript and the BoneScript API for interfacing Bea-gleBone Black to the physical world. BoneScript provides an easy way to interface the Bone to the world, but there are other programming languages out there. This recipe provides the steps needed to work with Python, and Recipe 5.21 does the same for C.

Adafruit (*http://www.adafruit.com/*) has produced an excellent library for the Bone (*http://bit.ly/1Mt9kDS*), and it's already installed. Ensure that the BBIO library is up-to-date:

```
bone# pip install Adafruit_BBIO
Requirement already satisfied (use --upgrade to upgrade):
    Adafruit-BBIO in /usr/local/lib/python2.7/dist-packages
Cleaning up...
```

Now that you're ready to go, add the code in Example 5-3 to a file called *blinkLED.py*.

Example 5-3. Use Python to blink an LED (blinkLED.py)

```
#!/usr/bin/env python
import Adafruit_BBIO.GPIO as GPIO
import time

pin = "P9_14"

GPIO.setup(pin, GPIO.OUT)

while True:
    GPIO.output(pin, GPIO.HIGH)
    time.sleep(0.5)
    GPIO.output(pin, GPIO.LOW)
    time.sleep(0.5)
```

Wire your LED as shown in Recipe 3.2 and run it with the following command:

```
bone# chmod +x blinkLED.py.
bone# ./blinkLED.py.
```

Adafruit's Setting up IO Python Library on BeagleBone Black (*http://bit.ly/1C5dkaK*) does a nice job showing how to use the BBIO library.

Discussion

Many of the constructs we have been using with BoneScript carry right over to Python. The BBIO GitHub site (*http://bit.ly/1Mt9kDS*) provides examples of how to poll for inputs, use pulse width modulation (PWM), and so on.

If you comment-out the time.sleep(0.5), you can measure how fast Python can toggle the output pin.

5.21 Using C to Interact with the Physical World

Problem

You want to use C on the Bone to talk to the world.

Solution

The C solution isn't as simple as the BoneScript or Python solution (Recipe 5.20), but it does work and is much faster. libsoc (*https://github.com/jackmitch/libsoc*) is a "C library for interfacing with common SoC {system on chip} peripherals through generic kernel interfaces" by Jack Mitchell (*http://www.embed.me.uk/*). Here's how to get and install it:

```
bone# git clone https://github.com/jackmitch/libsoc.git
bone# cd libsoc
bone# ./autogen.sh      # Takes about a minute
bone# ./configure       # Another  minute
bone# make              # 30 seconds
bone# make install
```

This installs the libsoc library in */usr/local/lib*. Add the code in Example 5-4 to a file called *blinkLED.c*.

Example 5-4. Use C to blink an LED (blinkLED.c)

```c
#include <stdio.h>
#include <stdlib.h>
#include <unistd.h>
#include <sys/wait.h>

#include "libsoc_gpio.h"
#include "libsoc_debug.h"

// Blinks an LED attached to P9_14 (gpio1_18, 32+18=50)
#define GPIO_OUTPUT  50

int main(void) {
  gpio *gpio_output;     // Create gpio pointer
  libsoc_set_debug(1);   // Enable debug output
  // Request gpio
  gpio_output = libsoc_gpio_request(GPIO_OUTPUT, LS_SHARED);
  // Set direction to OUTPUT
  libsoc_gpio_set_direction(gpio_output, OUTPUT);

  libsoc_set_debug(0);    // Turn off debug printing for fast toggle

  int i;
  for (i=0; i<1000000; i++) {     // Toggle the GPIO 100 times
    libsoc_gpio_set_level(gpio_output, HIGH);
    // usleep(100000);            // sleep 100,000 uS
    libsoc_gpio_set_level(gpio_output, LOW);
    // usleep(100000);
  }

  if (gpio_output) {
    libsoc_gpio_free(gpio_output);  // Free gpio request memory
  }

  return EXIT_SUCCESS;
}
```

In BoneScript and Python, we refer to the GPIO pins by the header position (P9_14, for example). In libsoc, we refer to them by an internal GPIO number. Figure 5-18 shows how to map from header position to GPIO pin. For example, P9_14 has GPIO_50 next to it. That means libsoc needs to use GPIO 50 to interface with P9_14.

P9					P8			
DGND	1	2	DGND		DGND	1	2	DGND
VDD_3V3	3	4	VDD_3V3		GPIO_38	3	4	GPIO_39
VDD_5V	5	6	VDD_5V		GPIO_34	5	6	GPIO_35
SYS_5V	7	8	SYS_5V		GPIO_66	7	8	GPIO_67
PWR_BUT	9	10	SYS_RESETN		GPIO_69	9	10	GPIO_68
GPIO_30	11	12	GPIO_60		GPIO_45	11	12	GPIO_44
GPIO_31	13	14	GPIO_50		GPIO_23	13	14	GPIO_26
GPIO_48	15	16	GPIO_51		GPIO_47	15	16	GPIO_46
GPIO_5	17	18	GPIO_4		GPIO_27	17	18	GPIO_65
	19	20			GPIO_22	19	20	GPIO_63
GPIO_3	21	22	GPIO_2		GPIO_62	21	22	GPIO_37
GPIO_49	23	24	GPIO_15		GPIO_36	23	24	GPIO_33
GPIO_117	25	26	GPIO_14		GPIO_32	25	26	GPIO_61
GPIO_115	27	28	GPIO_113		GPIO_86	27	28	GPIO_88
GPIO_111	29	30	GPIO_112		GPIO_87	29	30	GPIO_89
GPIO_110	31	32	VDD_ADC		GPIO_10	31	32	GPIO_11
AIN4	33	34	GNDA_ADC		GPIO_9	33	34	GPIO_81
AIN6	35	36	AIN5		GPIO_8	35	36	GPIO_80
AIN2	37	38	AIN3		GPIO_78	37	38	GPIO_79
AIN0	39	40	AIN1		GPIO_76	39	40	GPIO_77
GPIO_20	41	42	GPIO_7		GPIO_74	41	42	GPIO_75
DGND	43	44	DGND		GPIO_72	43	44	GPIO_73
DGND	45	46	DGND		GPIO_70	45	46	GPIO_71

Figure 5-18. Mapping from header pin to internal GPIO number

Compile and run the code:

```
bone# gcc -o blinkLED blinkLED.c -lsoc
bone# ./blinkLED
libsoc-debug: debug enabled (libsoc_set_debug)
libsoc-gpio-debug: requested gpio (50, libsoc_gpio_request)
libsoc-gpio-debug: GPIO already exported (50, libsoc_gpio_request)
libsoc-gpio-debug: setting direction to out (50, libsoc_gpio_set_direction)
libsoc-debug: debug disabled (libsoc_set_debug)
```

If you have an older image, you might need to install the libsoc library. You can do that with apt-get if you have network access and the right feeds on your Bone. Install libsoc2 for the library and libsoc-dev for the header files.

Discussion

It takes a little more code than Python, but it works, and it can be fast. Try uncommenting `libsoc_set_debug(0)` and commenting-out the `usleep`s to see how fast it can go. I measured a 6 µs period without the sleeps.

The test folder on GitHub (*http://bit.ly/1EXDFYO*) has some good sample code that shows how to set up callback to respond to changes in an input pin's value, `i2c`, `pwm`, `spi`, and more.

## CHAPTER 6

Internet of Things

6.0 Introduction

You can easily connect BeagleBone Black to the Internet via a wire (Recipe 5.11), wirelessly (Recipe 5.12), or through the USB to a host and then to the Internet (Recipe 5.13). Either way, it opens up a world of possibilities for the "Internet of Things" (IoT).

Now that you're online, this chapter offers various things to do with your connection.

6.1 Accessing Your Host Computer's Files on the Bone

Problem

You want to access a file on a Linux host computer that's attached to the Bone.

Solution

If you are running Linux on a host computer attached to BeagleBone Black, it's not hard to mount the Bone's files on the host or the host's files on the Bone by using sshfs. Suppose that you want to access files on the host from the Bone. First, install sshfs:

```
bone# apt-get install sshfs
```

Now, mount the files to an empty directory (substitute your username on the host computer for username and the IP address of the host for 192.168.7.1):

```
bone# mkdir host
bone# sshfs username@$192.168.7.1:. host
bone# cd host
bone# ls
```

The `ls` command will now list the files in your home directory on your host computer. You can edit them as if they were local to the Bone. You can access all the files by substituting `:/` for the `:.` following the IP address.

You can go the other way, too. Suppose that you are on your Linux host computer and want to access files on your Bone. Install `sshfs`:

```
host$ sudo apt-get install sshfs
```

and then access:

```
host$ mkdir /mnt/bone
host$ sshfs root@$192.168.7.2:/ /mnt/bone
host$ cd /mnt/bone
host$ ls
```

Here, we are accessing the files on the Bone as `root`. We've mounted the entire file system, starting with `/`, so you can access any file. Of course, with great power comes great responsibility, so be careful.

Discussion

The `sshfs` command gives you easy access from one computer to another. When you are done, you can unmount the files by using the following commands:

```
host$ umount /mnt/bone
bone# umount home
```

6.2 Serving Web Pages from the Bone

Problem

You want to use BeagleBone Black as a web server.

Solution

BeagleBone Black already has a couple of web servers running. This recipe shows how to use the Node.js-based server. Recipe 6.13 shows how to use *Apache*.

When you point your browser to *192.168.7.2*, you are using the Node.js web server The web pages are served from */var/lib/cloud9*. Add the HTML in Example 6-1 to a file called */var/lib/cloud9/test.html*, and then point your browser to *192.168.7.2:test.html*.

Example 6-1. A sample web page (test.html)

```
<!DOCTYPE html>
<html>
<body>

<h1>My First Heading</h1>

<p>My first paragraph.</p>

</body>
</html>
```

You will see the web page shown in Figure 6-1.

Figure 6-1. test.html as served by Node.js

Discussion

The Bone is already set up to serve many pages from */var/lib/cloud9*. For example, */var/lib/cloud9/Support* has the help pages for BoneScript. If you are going to create many of your own pages, consider putting them in their own directory.

6.3 Interacting with the Bone via a Web Browser

Problem

BeagleBone Black is interacting with the physical world nicely via BoneScript, and you want to display that information on a web browser.

Solution

Node.js (*http://nodejs.org/*) is a platform built on Chrome's JavaScript runtime (*http://code.google.com/p/v8/*) for easily building fast, scalable network applications. Recipe 6.2 shows how to use the node server that's already running. This recipe shows how easy it is to build your own server.

First, create a directory for this recipe:

```
bone# mkdir serverExample
bone# cd serverExample
```

Add the code in Example 6-2 to a file called *server.js* in the *serverExample* directory.

Example 6-2. Javascript code for a simple Node.js-based web server (server.js)

```
#!/usr/bin/env node
// Initial idea from Getting Started With node.js and socket.io
// by Constantine Aaron Cois, Ph.D. (www.codehenge.net)
// http://codehenge.net/blog/2011/12/getting-started-with-node-js-and-socket-io-
//    v0-7-part-2/
// This is a simple server for the various web frontends
"use strict";

var port = 9090,           // Port on which to listen
    http = require('http'),
    url = require('url'),
    fs = require('fs'),
    b = require('bonescript');

var server = http.createServer(servePage);    ❶

server.listen(port);                          ❷
console.log("Listening on " + port);

function servePage(req, res) {
    var path = url.parse(req.url).pathname;               ❸
    console.log("path: " + path);

    fs.readFile(__dirname + path, function (err, data) {❹
        if (err) {                                       ❺
            return send404(res);
        }
```

```
        res.write(data, 'utf8');                    ❻
        res.end();
    });
}

function send404(res) {
    res.writeHead(404);
    res.write('404 - page not found');
    res.end();
}
```

❶ Creates the server. Whenever a browser requests a page from port 9090, the func-
    tion servepage is called.

❷ Instructs the server to begin listening on port 9090.

❸ When servepage is called, req contains the path to the page being requested.
    This code extracts the path to the page.

❹ readFile reads the entire page being requested.

❺ Be sure the page is there; if not, send the famous 404 error.

❻ Write the page to the web browser.

Add the HTML in Example 6-1 to a file called *test.html* in the *serverExample* directory, and then run this:

```
bone# cd serverExample
bone# chmod +x server.js
bone# ./server.js
Listening on 9090
```

Now, navigate to *192.168.7.2:9090/test.html* on your host computer (or the Bone)
browser, and you will see something like Figure 6-2.

Figure 6-2. Test page served by our custom node-based server

Note that the only difference between Figure 6-2 and Figure 6-1 is that the first uses port 9090.

Discussion

You can add whatever files you want; just put them in the *serverExample* directory with *server.js*. The files can link to one another and external pages. It will all just work.

6.4 Displaying GPIO Status in a Web Browser

Problem

You want a web page to display the status of a GPIO pin.

Solution

This solution builds on the Node.js-based web server solution in Recipe 6.3.

To make this recipe, you will need:

- Breadboard and jumper wires (see "Prototyping Equipment" on page 316)
- Pushbutton switch (see "Miscellaneous" on page 318)

Wire your pushbutton as shown in Figure 2-5. Add the code in Example 6-3 to a file called *GPIOserver.js* in the *serverExample* directory.

Example 6-3. A simple Node.js-based web server to read a GPIO (GPIOserver.js)

```
#!/usr/bin/env node
// Initial idea from Getting Started With node.js and socket.io
// http://codehenge.net/blog/2011/12/getting-started-with-node-js-and-socket-io-
//    v0-7-part-2/
// This is a simple server for the various web frontends
// Display status of P9_42
"use strict";

var port = 9090,            // Port on which to listen
    http = require('http'),
    url = require('url'),
    fs = require('fs'),
    b = require('bonescript'),
    gpio = 'P9_42';         // gpio port to read
                ❶
var htmlStart = "\
<!DOCTYPE html>\
<html>\
<body>\
\
<h1>" + gpio + "</h1>\
data = ";
                ❷
var htmlEnd = "\
</body>\
</html>";

var server = http.createServer(servePage);

b.pinMode(gpio, b.INPUT, 7, 'pulldown');

server.listen(port);
console.log("Listening on " + port);

function servePage(req, res) {
    var path = url.parse(req.url).pathname;
    console.log("path: " + path);
    if (path === '/gpio') {                         ❸
        var data = b.digitalRead(gpio);             ❹
        res.write(htmlStart + data + htmlEnd, 'utf8');   ❺
```

```
            res.end();
        } else {
            fs.readFile(__dirname + path, function (err, data) {
                if (err) {
                    return send404(res);
                }
                res.write(data, 'utf8');
                res.end();
            });
        }
}

function send404(res) {
    res.writeHead(404);
    res.write('404 - page not found');
    res.end();
}
```

❶ This the the HTML to send to the server before the data value.

❷ This is the HTML for after the data.

❸ Normally, we'd send a file, but if /gpio is requested, we'll build our own page and include the value of the GPIO pin.

❹ Read the value of the GPIO pin.

❺ Concatenate the HTML code with the data and send it to the browser.

Now, run the following command:

```
bone# ./GPIOserver.js
Listening on 9090
```

Point your browser to *http://192.168.7.2:9090/gpio*, and the page will look like Figure 6-3.

Figure 6-3. Status of a GPIO pin on a web page

Currently, the 0 shows that the button isn't pressed. Try refreshing the page while pushing the button, and you will see 1 displayed.

Discussion

It's not hard to assemble your own HTML with the GPIO data. It's an easy extension to write a program to display the status of all the GPIO pins.

6.5 Continuously Displaying the GPIO Value via jsfiddle

Problem

The value of your GPIO port keeps changing, and you want your web page to always show the current value.

Solution

We'll attack this problem in two steps. This recipe will show how to use an external website (jsfiddle [*http://jsfiddle.net/*]) to read a GPIO pin on the Bone (yup, an external site can read the GPIO pins on the Bone). Recipe 6.6 will show how to move the code to run locally on the Bone.

To make this recipe, you will need:

- Breadboard and jumper wires (see "Prototyping Equipment" on page 316)
- Pushbutton switch (see "Miscellaneous" on page 318)
- 10 kΩ trimpot (variable resistor) (see "Resistors" on page 316)

jsfiddle (*http://jsfiddle.net/*) is a website that lets you fiddle with JavaScript. It's best introduced by an example. Point your browser to jQuery example (*http://bit.ly/ 19c75ql*), and you will see the screen shown in Figure 6-4.

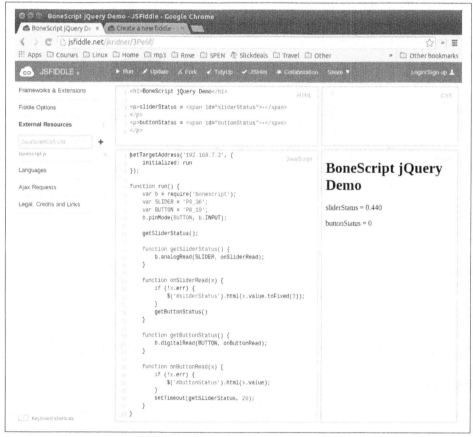

Figure 6-4. jsfiddle example of GPIO

The page has four panels. The upper-left panel shows your HTML code. The lower-left panel shows the JavaScript that animates the HTML. The upper-right panel shows the CSS code that formats your HTML (not used here), and the lower-right panel shows what your rendered page looks like.

Wire your pushbutton switch to P8_19 (use Figure 2-5 as an example, but note that the pushbutton in that figure is wired to P9_42, instead) and the variable resistor to P9_36, as shown in Figure 2-8.

You can use any GPIO pin and any analog-in pin; just edit the code to match your setup.

Press the Run button on the jsfiddle page. The values in the lower-right panel will track your button and variable resistor in real time.

Discussion

Wow! What is happening here? The JavaScript code is running on the browser, but when you get to a BoneScript statement such as b.analogRead(SLIDER, onSlider Read);, the browser does a remote procedure call (RPC) to the Bone to read the value. The value is returned to the browser, and it continues running. This is a very powerful setup. The same BoneScript code that runs on the Bone and interacts with the hardware can also run on a remote browser and *still interact with the Bone!*

Let's take a close look at the code in Example 6-4.

Example 6-4. BoneScript running in a browser and reading BeagleBone Black hardware (jQueryDemo.js)

```
setTargetAddress('192.168.7.2', {          ❶
    initialized: run
});

function run() {
    var b = require('bonescript');         ❷
    var SLIDER = 'P9_36';                   ❸
    var BUTTON = 'P8_19';
    b.pinMode(BUTTON, b.INPUT);

    getSliderStatus();                      ❹

    function getSliderStatus() {
        b.analogRead(SLIDER, onSliderRead); ❺
    }

    function onSliderRead(x) {
        if (!x.err) {                       ❻
```

```
        $('#sliderStatus').html(x.value.toFixed(3));
      }
      getButtonStatus()                          ❼
    }

    function getButtonStatus() {
      b.digitalRead(BUTTON, onButtonRead);       ❽
    }

    function onButtonRead(x) {
      if (!x.err) {                              ❾
        $('#buttonStatus').html(x.value);
      }
      setTimeout(getSliderStatus, 20);           ❿
    }
  }
```

❶ Direct the browser where to find your Bone. After the connection is made, run is called.

❷ Bring in the BoneScript library.

❸ Select which analog port (SLIDER) and which GPIO pin (BUTTON) to use.

❹ Call getSliderStatus() to get things started.

❺ Read the analog-in and call onSliderRead when the value is available.

❻ Here's an important step. If there is no error, set the value of sliderStatus to the value read. x.value is the analog value read. Adding .toFixed(3) makes it display only three digits. $('#sliderStatus') is our connection to the HTML. The # specifies to look for the HTML with the matching ID. In our case, $('slider Status') instructs us to look for the HTML elements with id="sliderStatus". In this example, it's <span id="sliderStatus">-</span>. Adding the .html() (giving $('#sliderStatus').html) makes the statement refer to everything between <span> and </span>, which initially is just -. The entire statement says to change whatever is between the spans to the analog value that was read. And the web page instantly changes that value.

❼ Now, move on to reading the button.

❽ As before, call onButtonRead when the value is read.

❾ If no error, update the element with id="buttonStatus".

❿ Finally, wait 20 ms and start all over again.

We've just opened up a whole now world with the Bone. Now, you can easily grab data from the physical world via sensors on the Bone and display it in a browser anywhere in the world.

See Also

jsfiddle is great for experimenting with your code, but what if want everything to be served from your Bone and not use an external site? Check out Recipe 6.6.

6.6 Continuously Displaying the GPIO Value

Problem

The value of your GPIO port keeps changing, and you want your web page to always show the current value, but you don't want to use an external site.

Solution

Recipe 6.5 shows how to use BoneScript in a browser to read GPIO pins on the Bone, but it requires the use of an external website. This recipe shows how to move from jsfiddle to the Bone, with a few simple steps.

Go to a working jsfiddle page (such as the one shown in Figure 6-4) and do the following:

1. In the lower-left panel, select all the BoneScript code and copy it to a file. In this case, call it *jQueryDemo.js*. Save it in the *serverExample* directory.

2. From the upper-left panel, select all the HTML code and copy it to *jQueryDemo.html* in the *serverExample* directory.

3. We need to add code to inform the browser where to find the external libraries we are using. Edit *jQueryDemo.html* and *before* your HTML, add the following lines:

```
<html>
    <head>
        <title>BoneScript jQuery Demo</title>
        <script src="static/jquery.js"></script>
        <script src="static/bonescript.js"></script>
        <script src="jQueryDemo.js"></script>
    </head>
<body>
```

4. Add the following lines *after* your HTML:

```
</body>
</html>
```

5. Open *http://192.168.7.2/serverExample/jQueryDemo.html*, and you'll now see your values in your own web page.

 There are several useful JavaScript libraries already located in */var/lib/cloud9/static*, and this recipe makes use of a couple of them here.

Discussion

The additional HTML used in this recipe is loaded in code for *jquery.js*, *bonescript.js*, and our own *jQueryDemo.js*. How did we know to add *jquery.js* and *bonescript.js*? Look in the left column of Figure 6-4. Under External Resources, *bonescript.js* is listed, thus alerting us that we need to include it. But what about *jquery.js*? Click Frameworks & Extensions, and you'll see Figure 6-5.

Figure 6-5. jsfiddle Frameworks & Extensions

Here, jQuery is listed. This lets us know that we need to include it.

6.7 Plotting Data

Problem

1. You have live data coming into your Bone, and you want to plot it.

Solution

There are a number of JavaScript-based plotting packages (*http://bit.ly/1MrHm9L*) out there. This recipe uses flot (*http://www.flotcharts.org/*) because it strikes a nice balance between ease-of-use and power.

To make this recipe, you will need:

- Breadboard and jumper wires (see "Prototyping Equipment" on page 316)
- 10 kΩ trimpot (variable resistor) (see "Resistors" on page 316)

Wire your variable resistor to P9_36, as shown in Figure 2-8. Go to jsfiddle Flot Demo (*http://bit.ly/1E5DW85*) for a nice quick demo. You will the screen shown in Figure 6-6. Try changing the variable resistor, and watch the plot respond in real time.

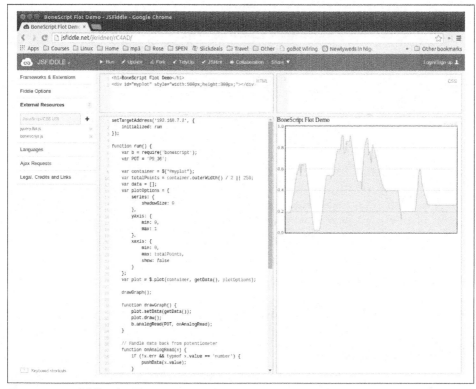

Figure 6-6. jsfiddle Flot Demo

See Recipe 6.5 for a description of the jsfiddle panels. In Figure 6-6, the lower-right panel displays a live scrolling chart of the values being read from the variable resistor.

Discussion

You can perform plotting externally, via the jsfiddle (*http://jsfiddle.net/*) site, or locally.

Plotting with jsfiddle

The HTML panel in Figure 6-6 simply creates an area called myplot in which to draw the plot. The JavaScript code in Example 6-5 then reads the analog in and creates the plot to display.

Example 6-5. Code for plotting data with Flot (flotDemo.js)

```
setTargetAddress('192.168.7.2', {                              ❶
    initialized: run
});

function run() {
    var b = require('bonescript');                             ❷
    var POT = 'P9_36';                                         ❸

    var container = $("#myplot");                              ❹
    var totalPoints = container.outerWidth() / 2 || 250;       ❺
    var data = [];
    var plotOptions = {                                        ❻
        series: {
            shadowSize: 0
        },
        yaxis: {
            min: 0,
            max: 1
        },
        xaxis: {
            min: 0,
            max: totalPoints,
            show: false
        }
    };
    var plot = $.plot(container, getData(), plotOptions);      ❼

    drawGraph();                                               ❽

    function drawGraph() {                                     ❾
        plot.setData(getData());
        plot.draw();
        b.analogRead(POT, onAnalogRead);
    }

    // Handle data back from potentiometer
    function onAnalogRead(x) {
        if (!x.err && typeof x.value == 'number') {
            pushData(x.value);                                 ❿
        }
        setTimeout(drawGraph, 20);                             ⓫
    }
```

```
function pushData(y) {
    if (data.length && (data.length + 1) > totalPoints) {
        data = data.slice(1);
    }
    if (data.length < totalPoints) {
        data.push(y);
    }
}

function getData() {
    var res = [];
    for (var i = 0; i < data.length; ++i) {
        res.push([i, data[i]]);
    }
    var series = [{
        data: res,
        lines: {
            fill: true
        }
    }];
    return series;
}
}
```

❶ Direct the browser where to find the Bone. After it is initialized, it calls run to get things going.

❷ Read in the BoneScript library.

❸ Pick which analog-in to use.

❹ Pick where on the web page to draw the plot. In this case, the code defines only one place, myplot.

❺ Decide how many points can be plotted.

❻ Set various plotting options.

❼ Initialize the plot.

❽ Draw the graph on the web page.

❾ Set the data and draw. Then read the next value and call onAnalogRead when it's ready.

❿ If no error, add newest value to the data array.

⓫ Do it all over again in 20 ms.

Plotting locally

You can do plotting locally by following the process in Recipe 6.6:

1. Copy the JavaScript code in the lower-left panel and save it in a file called *flotDemo.js*.

2. Copy the HTML in the upper-left panel and add the header material. Example 6-6 loads the `flot` library in addition to `jquery` and our own code.

Example 6-6. flotDemo.html (flotDemo.html)

```html
<html>
    <head>
        <title>BoneScript Flot Demo</title>
        <script src="/static/flot/jquery.min.js"></script>
        <script src="/static/flot/jquery.flot.min.js"></script>
        <script src="/static/bonescript.js"></script>
        <script src="flotDemo.js"></script>
    </head>
<body>

<h1>BoneScript Flot Demo</h1>
<div id="myplot" style="width:500px;height:300px;"></div>

</body>
</html>
```

3. Put both files in the *serverExample* subdirectory under the Cloud9 IDE (*/var/lib/cloud9*).

4. Browse to *http://192.168.7.2/serverExample/flotDemo.html*, and you'll have a local version of your demo running.

Go explore the Flot site (*http://www.flotcharts.org/*) and see what other plots you can do.

6.8 Sending an Email

Problem

You want to send an email via Gmail from the Bone.

Solution

First, you need to set up a Gmail account (*https://mail.google.com*), if you don't already have one. Next, install `nodemailer`:

```
bone# npm install -g nodemailer
```

Then add the code in Example 6-7 to a file named *nodemailer-test.js* in the *serverExample* directory. Be sure to substitute your own Gmail username and password for user and pass.

Example 6-7. Sending email using nodemailer (nodemailer-test.js)

```
#!/usr/bin/env node
// From: https://github.com/andris9/Nodemailer

var nodemailer = require('nodemailer');

// create reusable transporter object using SMTP transport
var transporter = nodemailer.createTransport({
    service: 'gmail',
    auth: {
        user: 'yourUser@gmail.com',
        pass: 'yourPass'
    }
});

// NB! No need to re-create the transporter object. You can use
// the same transporter object for all e-mails

// set up e-mail data with unicode symbols
var mailOptions = {
    from: 'Your User <yourUser@gmail.edu>', // sender address
    to: 'anotherUser@gmail.edu', // list of receivers
    subject: 'Test of nodemail', // Subject line
    text: 'Hello world from modemailer', // plaintext body
    html: '<b>Hello world</b><p>Way to go!</p>' // html body
};

// send mail with defined transport object
transporter.sendMail(mailOptions, function(error, info){
    if(error){
        console.log(error);
    }else{
        console.log('Message sent: ' + info.response);
    }
});

// Nodemailer is licensed under MIT license
// (https://github.com/andris9/Nodemailer/blob/master/LICENSE).
// Basically you can do whatever you want to with it
```

Then run the script to send the email:

```
bone# chmod +x nodemailer-test.js
bone# .\nodemailer-test.js
```

This solution requires your Gmail password to be in plain text in a file, which is a security problem. Make sure you know who has access to your Bone. Also, if you remove the microSD card, make sure you know who has access to it. Anyone with your microSD card can read your Gmail password.

Discussion

Be careful about putting this into a loop. Gmail presently limits you to 500 emails per day and 10 MB per message (*http://group-mail.com/email-marketing/how-to-send-bulk-emails-using-gmail/*).

6.9 Sending an SMS Message

Problem

You want to send a text message from BeagleBone Black.

Solution

There are a number of SMS services out there. This recipe uses Twilio because you can use it for free, but you will need to verify the number (*http://bit.ly/1MrHBBF*) to which you are texting. First, go to Twilio's home page (*https://www.twilio.com/*) and set up an account. Note your account SID and authorization token. If you are using the free version, be sure to verify your numbers (*http://bit.ly/19c7GZ7*).

Next, install Trilio by using the following command:

```
bone# npm install -g twilio
```

Finally, add the code in Example 6-8 to a file named *twilio-test.js* and run it. Your text will be sent.

Example 6-8. Sending SMS messages using Twilio (twilio-test.js)

```
#!/usr/bin/env node
// From: http://twilio.github.io/twilio-node/
// Twilio Credentials
var accountSid = '';
var authToken = '';

//require the Twilio module and create a REST client
var client = require('twilio')(accountSid, authToken);

client.messages.create({
        to: "812555121",
        from: "+2605551212",
        body: "This is a test",
```

```
}, function(err, message) {
      console.log(message.sid);
});

// https://github.com/twilio/twilio-node/blob/master/LICENSE
// The MIT License (MIT)
// Copyright (c) 2010 Stephen Walters
// Copyright (c) 2012 Twilio Inc.

// Permission is hereby granted, free of charge, to any person obtaining a copy of
// this software and associated documentation files (the "Software"), to deal in
// the Software without restriction, including without limitation the rights to
// use, copy, modify, merge, publish, distribute, sublicense, and/or sell copies
// of the Software, and to permit persons to whom the Software is furnished to do
// so, subject to the following conditions:

// The above copyright notice and this permission notice shall be included in
// all copies or substantial portions of the Software.

// THE SOFTWARE IS PROVIDED "AS IS", WITHOUT WARRANTY OF ANY KIND, EXPRESS OR
// IMPLIED, INCLUDING BUT NOT LIMITED TO THE WARRANTIES OF MERCHANTABILITY,
// FITNESS FOR A PARTICULAR PURPOSE AND NONINFRINGEMENT. IN NO EVENT SHALL
// THE AUTHORS OR COPYRIGHT HOLDERS BE LIABLE FOR ANY CLAIM, DAMAGES OR OTHER
// LIABILITY, WHETHER IN AN ACTION OF CONTRACT, TORT OR OTHERWISE, ARISING
// FROM, OUT OF OR IN CONNECTION WITH THE SOFTWARE OR THE USE OR OTHER
// DEALINGS IN THE SOFTWARE.
```

Discussion

Twilio allows a small number of free text messages, enough to test your code and to play around some.

6.10 Displaying the Current Weather Conditions

Problem

You want to display the current weather conditions.

Solution

Because your Bone is on the network, it's not hard to access the current weather conditions from a weather API. First, install a JavaScript module:

```
bone# npm install -g weather-js   (Took about 4 minutes on a slow connection)
```

Then add the code in Example 6-9 to a file named *weather.js*.

Example 6-9. Code for getting current weather conditions (weather.js)

```
#!/usr/bin/env node
// Install with npm install weather-js
var weather = require('weather-js');

// Options:
// search:    location name or zipcode
// degreeType: F or C

weather.find({search: 'Terre Haute, IN', degreeType: 'F'},
  function(err, result) {
    if (err) {
      console.log(err);
    }
    console.log(JSON.stringify(result, null, 2));                  ❶
    console.log(JSON.stringify(result[0].current, null, 2));       ❷
    console.log(JSON.stringify(result[0].forecast[0], null, 2));   ❸
  });
```

❶ Prints everything returned by the weather site.

❷ Prints only the current weather conditions.

❸ Prints the forecast for the next day.

Run this by using the following commands:

```
bone# chmod +x weather.js
bone# ./weather.js
...
{
  "temperature": "74",
  "skycode": "28",
  "skytext": "Mostly Cloudy",
  "date": "2014-07-28",
  "observationtime": "14:53:00",
  "observationpoint": "Terre Haute, Terre Haute International Airport -
      Hulman Field",
  "feelslike": "74",
  "humidity": "53",
  "winddisplay": "15 mph NW",
  "day": "Monday",
  "shortday": "Mon",
  "windspeed": "15",
  "imageUrl": "http://wst.s-msn.com/i/en-us/law/28.gif"
}
```

Discussion

The weather API returns lots of information. Use JavaScript to extract the information you want.

6.11 Sending and Receiving Tweets

Problem

You want to send and receive tweets (Twitter posts) with your Bone.

Solution

First, install the Twitter node package:

```
bone# npm install node-twitter
```

Before creating a Twitter app, you need to authenticate with Twitter:

1. Go to the Twitter Application Management page (*https://apps.twitter.com/*) and click Create New App.

2. Fill in the requested information and click "Create your Twitter application." You can leave the Callback URL blank.

3. Click the API Keys tab. If you plan to have the Bone send tweets, click "modify app permissions" to the right of "Access level."

4. Select "Read and Write" and click "Update settings."

5. Click the API Keys tab.

6. Near the bottom of the page, click "Create my access token."

7. Create a file called *twitterKeys.js* and add the following code to it, substituting your keys for *xxx*:

```
exports.API_KEY      = 'xxx';
exports.API_SECRET   = 'xxx';
exports.TOKEN        = 'xxx';
exports.TOKEN_SECRET = 'xxx';
```

Add the code in Example 6-10 to a file called *twitterTimeLine.js* and run it to see your timeline.

Example 6-10. Code to display your Twitter timeline (twitterTimeLine.js)

```
#!/usr/bin/env node
// From: https://www.npmjs.org/package/node-twitter
// The Twitter REST API can be accessed using Twitter.RestClient. The following
// code example shows how to retrieve tweets from the authenticated user's timeline.
```

```
var Twitter = require('node-twitter');
var key = require('./twitterKeys');

var twitterRestClient = new Twitter.RestClient(
    key.API_KEY, key.API_SECRET,
    key.TOKEN,   key.TOKEN_SECRET
);

twitterRestClient.statusesHomeTimeline({}, function(error, result) {
    if (error) {
        console.log('Error: ' +
            (error.code ? error.code + ' ' + error.message : error.message));
    }

    if (result) {
        console.log(result);
    }
});

// node-twitter is made available under terms of the BSD 3-Clause License.
// http://www.opensource.org/licenses/BSD-3-Clause
```

Use the code in Example 6-11 to send a tweet with a picture.

Example 6-11. Code to send a tweet with a picture (twitterUpload.js)

```
#!/usr/bin/env node
// From: https://www.npmjs.org/package/node-twitter
// Tweets with attached image media (JPG, PNG or GIF) can be posted
// using the upload API endpoint.
var Twitter = require('node-twitter');
var b = require('bonescript');
var key = require('./twitterKeys');

var twitterRestClient = new Twitter.RestClient(
    key.API_KEY, key.API_SECRET,
    key.TOKEN,   key.TOKEN_SECRET
);

twitterRestClient.statusesUpdateWithMedia(
    {
        'status': 'Posting a tweet w/ attached media.',
        'media[]': '/root/cookbook-atlas/images/cover.png'
    },
    function(error, result) {
        if (error) {
            console.log('Error: ' +
                (error.code ? error.code + ' ' + error.message : error.message));
        }

        if (result) {
            console.log(result);
```

```
            }
        }
);

// node-twitter is made available under terms of the BSD 3-Clause License.
// http://www.opensource.org/licenses/BSD-3-Clause
```

The code in Example 6-12 sends a tweet whenever a button is pushed.

Example 6-12. Tweet when a button is pushed (twitterPushbutton.js)

```
#!/usr/bin/env node
// From: https://www.npmjs.org/package/node-twitter
// Tweets with attached image media (JPG, PNG or GIF) can be posted
// using the upload API endpoint.
var Twitter = require('node-twitter');
var b = require('bonescript');
var key = require('./twitterKeys');
var gpio = "P9_42";
var count = 0;

b.pinMode(gpio, b.INPUT);
b.attachInterrupt(gpio, sendTweet, b.FALLING);

var twitterRestClient = new Twitter.RestClient(
    key.API_KEY, key.API_SECRET,
    key.TOKEN,   key.TOKEN_SECRET
);

function sendTweet() {
    console.log("Sending...");
    count++;

    twitterRestClient.statusesUpdate(
        {'status': 'Posting tweet ' + count + ' via my BeagleBone Black', },
        function(error, result) {
            if (error) {
                console.log('Error: ' +
                    (error.code ? error.code + ' ' + error.message : error.message));
            }

            if (result) {
                console.log(result);
            }
        }
    );
}

// node-twitter is made available under terms of the BSD 3-Clause License.
// http://www.opensource.org/licenses/BSD-3-Clause
```

To see many other examples, go to iStrategyLabs' node-twitter GitHub page (*http://bit.ly/18AvSTW*).

Discussion

This opens up many new possibilities. You can read a temperature sensor and tweet its value whenever it changes, or you can turn on an LED whenever a certain hashtag is used. What are you going to tweet?

6.12 Wiring the IoT with Node-RED

Problem

You want BeagleBone to interact with the Internet, but you want to program it graphically.

Solution

Node-RED (*http://nodered.org/*) is a visual tool for wiring the IoT. It makes it easy to turn on a light when a certain hashtag is tweeted, or spin a motor if the forecast is for hot weather.

Installing Node-RED

To install Node-RED, run the following commands:

```
bone# cd            # Change to home directory
bone# git clone https://github.com/node-red/node-red.git
bone# cd node-red/
bone# npm install --production    # almost 6 minutes
bone# cd nodes
bone# git clone https://github.com/node-red/node-red-nodes.git # 2 seconds
bone# cd ~/node-red
```

To run Node-RED, use the following commands:

```
bone# cd ~/node-red
bone# node red.js
Welcome to Node-RED
===================

18 Aug 16:31:43 - [red] Version: 0.8.1.git
18 Aug 16:31:43 - [red] Loading palette nodes
18 Aug 16:31:49 - [26-rawserial.js] Info : only really needed for
                   Windows boxes without serialport npm module installed.
18 Aug 16:31:56 - ----------------------------------------
18 Aug 16:31:56 - [red] Failed to register 44 node types
18 Aug 16:31:56 - [red] Run with -v for details
18 Aug 16:31:56 - ----------------------------------------
```

```
18 Aug 16:31:56 - [red] Server now running at http://127.0.0.1:1880/
18 Aug 16:31:56 - [red] Loading flows : flows_yoder-debian-bone.json
```

The second-to-last line informs you that Node-RED is listening on part 1880. Point your browser to *http://192.168.7.2:1880*, and you will see the screen shown in Figure 6-7.

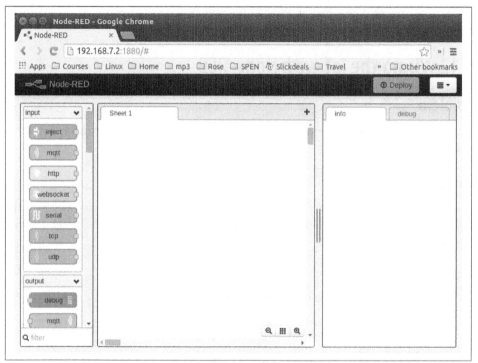

Figure 6-7. The Node-RED web page

Building a Node-RED Flow

The example in this recipe builds a Node-RED flow that will toggle an LED whenever a certain hashtag is tweeted. But first, you need to set up the Node-RED flow with the twitter node:

1. On the Node-RED web page, scroll down until you see the social nodes on the left side of the page.

2. Drag the twitter node to the canvas, as shown in Figure 6-8.

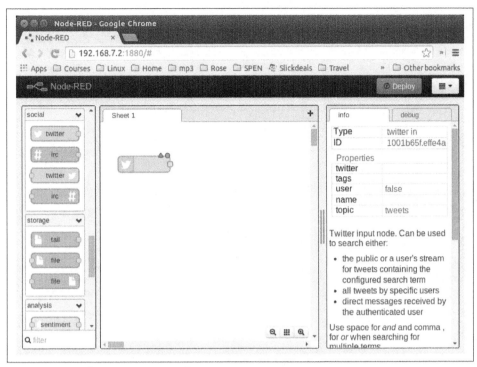

Figure 6-8. Node-RED twitter node

3. Authorize Twitter by double-clicking the `twitter` node. You'll see the screen shown in Figure 6-9.

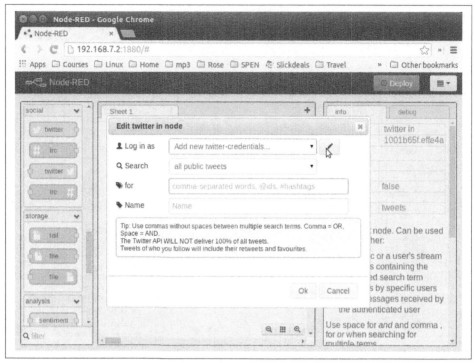

Figure 6-9. Node-RED Twitter authorization, step 1

4. Click the pencil button to bring up the dialog box shown in Figure 6-10.

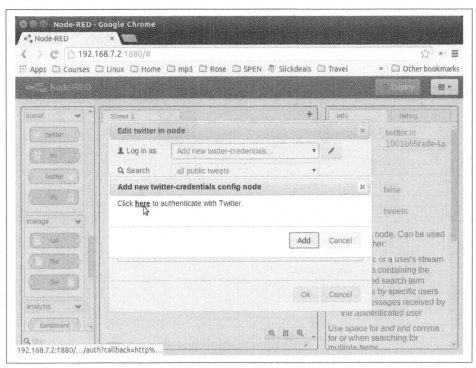

Figure 6-10. Node-RED twitter authorization, step 2

5. Click the "here" link, as shown in Figure 6-10, and you'll be taken to Twitter to authorize Node-RED.

6. Log in to Twitter and click the "Authorize app" button (Figure 6-11).

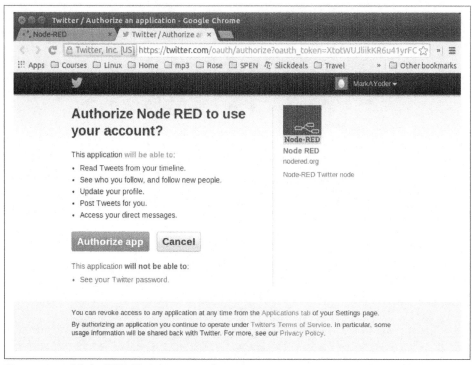

Figure 6-11. Node-RED Twitter site authorization

7. When you're back to Node-RED, click the Add button, add your Twitter credentials, enter the hashtags to respond to (Figure 6-12), and then click the Ok button.

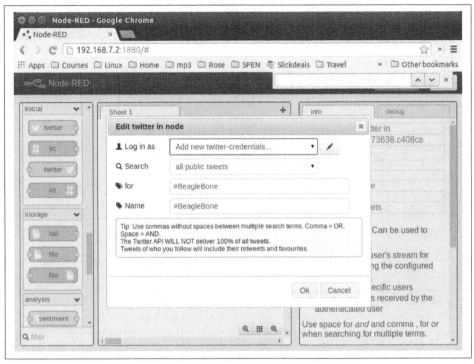

Figure 6-12. Node-RED adding the #BeagleBone hashtag

8. Go back to the left panel, scroll up to the top, and then drag the debug node to the canvas. (debug is in the output section.)

9. Connect the two nodes by clicking and dragging (Figure 6-13).

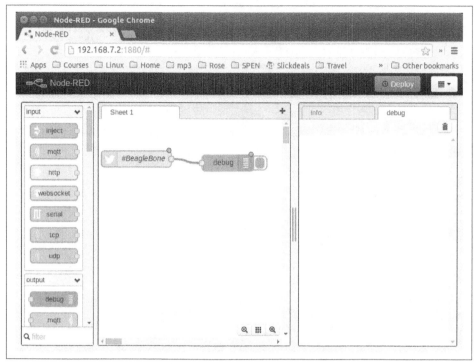

Figure 6-13. Node-RED Twitter adding debug node and connecting

10. In the right panel, in the upper-right corner, click the "debug" tab.

11. Finally, click the Deploy button above the "debug" tab.

Your Node-RED flow is now running on the Bone. Test it by going to Twitter and tweeting something with the hashtag #BeagleBone. Your Bone is now responding to events happening out in the world.

Adding an LED Toggle

Now, we're ready to add the LED toggle:

1. Wire up an LED as shown in Recipe 3.2. Mine is wired to P9_14.

2. Scroll to the bottom of the left panel and drag the bbb-discrete-out node (second from the bottom of the bbb nodes) to the canvas and wire it (Figure 6-14).

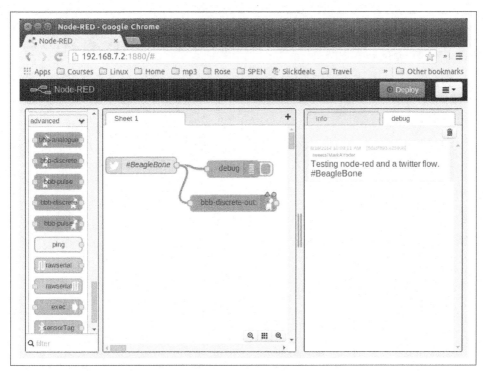

Figure 6-14. Node-RED adding bbb-discrete-out node

3. Double-click the node, select your GPIO pin and "Toggle state," and then set "Startup as" to 1 (Figure 6-15).

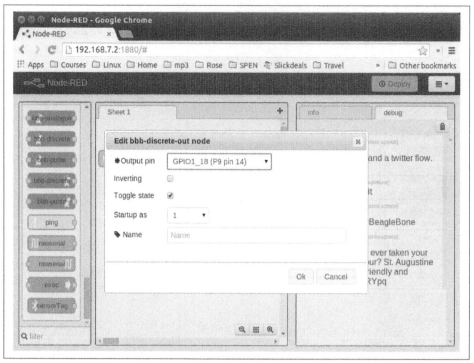

Figure 6-15. Node-RED adding bbb-discrete-out configuration

4. Click Ok and then Deploy.

Test again. The LED will toggle every time the hashtag #BeagleBone is tweeted. With a little more exploring, you should be able to have your Bone ringing a bell or spinning a motor in response to tweets.

Discussion

We've just opened up a whole other world of IoT. Go explore. You'll find all sorts of ways to interact with the Bone and the Internet. For example, if you install the following nodes, you can use Texas Instruments' SensorTag (Recipe 2.11) with Node-RED (be sure to restart Node-RED after installing):

```
bone# apt-get install libbluetooth-dev
bone# npm install -g sensortag
```

In fact, there are many other nodes that you can install, including these useful nodes:

```
npm install -g badwords ntwitter oauth sentiment wordpos xml2js firmata
npm install -g fs.notify serialport feedparser pushbullet irc simple-xmpp
npm install -g redis mongodb node-stringprep
npm install -g imap js2xmlparser nodemailer arduino-firmata heatmiser wemo
```

```
npm install -g stomp-client wake_on_lan node-dweetio komponist nma growl
npm install -g node-prowl pusher pushover-notifications snapchat twilio aws-sdk
npm install -g node-hue-api level pusher-client mysql sqlite3 suncalc
npm install -g dynamodb-data-types node-postgres-named
```

The Node-RED documentation (*http://nodered.org/docs/*) shows many other things you can do with Node-RED, and the Node-RED flows (*http://flows.nodered.org/*) site lists many things others are trying. Go and explore.

6.13 Serving Web Pages from the Bone by Using Apache

Problem

You want to use BeagleBone Black as an Apache web server.

Solution

Apache (*https://httpd.apache.org/*) (the most popular web server on the Internet since 1996) is already running on the Bone. Point your browser to *http://192.168.7.2:8080/*, and you'll see a screen that looks like Figure 6-16.

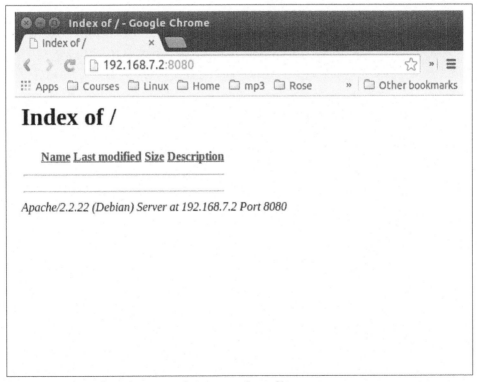

Figure 6-16. Apache running on the Bone with no files to view

What you are seeing is a listing of the files in Apache's root directory (*/var/www/*). In this case, there are no files to show. Add the HTML in Example 6-1 to a file called */var/www/test.html* and refresh your browser. You will see the screen shown in Figure 6-17.

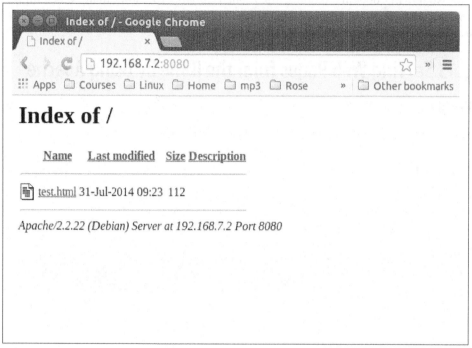

Figure 6-17. The test.html file appears

Click the *test.html* link to display the screen in Figure 6-18.

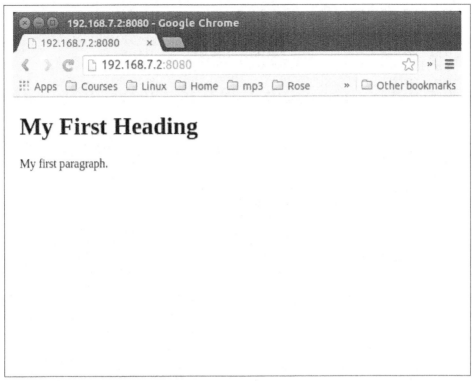

Figure 6-18. Apache running on the Bone showing test.html

You can place any number of files in */var/www/* and access them with Apache.

Discussion

A computer can have a number of services (such as SSH, FTP, or a web browser) that use the network. *Port* numbers are the way of specifying the computer to which service you want to connect. When you point your web browser to an IP address such as *192.168.7.2*, the browser assumes you want port 80. Apache defaults to listening on port 80, so your browser connects to Apache by default.

The Bone, however, has something else already listening to port 80, so Apache (on the Bone) has been configured to listen to port 8080. How do you know what port Apache is using? Try the following command:

```
bone# netstat -taupn
Active Internet connections (servers and established)
Proto Local Address         Foreign Address        PID/Program name
tcp   127.0.0.1:3350         0.0.0.0:*              880/xrdp-sesman
tcp   0.0.0.0:22             0.0.0.0:*              860/sshd
tcp   127.0.0.1:6010         0.0.0.0:*              1143/0
```

```
tcp     127.0.0.1:6011          0.0.0.0:*               17227/4
tcp     0.0.0.0:1883            0.0.0.0:*               625/mosquitto
tcp     0.0.0.0:3389            0.0.0.0:*               859/xrdp
tcp     192.168.7.2:22          192.168.7.1:59284       1143/0
tcp     192.168.7.2:22          192.168.7.1:45655       17227/4
tcp6    :::22                   :::*                    860/sshd
tcp6    :::3000                 :::*                    1/systemd
tcp6    ::1:6010                :::*                    1143/0
tcp6    ::1:6011                :::*                    17227/4
tcp6    :::1883                 :::*                    625/mosquitto
tcp6    :::8080                 :::*                    368/apache2
tcp6    :::80                   :::*                    1/systemd
tcp6    192.168.7.2:80          192.168.7.1:34251       1102/node
tcp6    192.168.7.2:3000        137.112.41.137:53894    1104/node
tcp6    192.168.7.2:3000        192.168.7.1:51422       1104/node
tcp6    192.168.7.2:3000        137.112.41.137:54480    1104/node
udp     0.0.0.0:58536           0.0.0.0:*               530/avahi-daemon: r
udp     0 0.0.0.0:5353           0.0.0.0:*               530/avahi-daemon: r
udp     0 0.0.0.0:67             0.0.0.0:*               1011/udhcpd
udp6    :::42706                :::*                    530/avahi-daemon: r
udp6    :::5353                 :::*                    530/avahi-daemon: r
```

 To allow the output to fit on the width of the printed page, the Recv-Q, Send-Q, and State columns have been removed from the preceding code listing.

The Local Address column lists the IP addresses and port numbers. Here, apache2 is shown listening on port 8080, and node is listening on port 80. Here's how to switch *Apache* back to port 80.

First, move node to another port by editing /lib/systemd/system/bonescript.socket. Initially, it begins with this:

```
[Socket]
ListenStream=80
```

Change the 80 to another port number—8081, for example.

Next, move apache2 to another port by editing the */etc/apache2/ports.conf* file. A few lines down, you will find these lines:

```
NameVirtualHost *:8080
Listen 8080
```

Change both instances of 8080 to 80 and reboot to make the changes effective.

Now, pointing a browser to *192.168.7.2* will use the Apache server.

6.14 Communicating over a Serial Connection to an Arduino or LaunchPad

Problem

You would like your Bone to talk to an Arduino or LaunchPad.

Solution

The common serial port (also know as a UART) is the simplest way to talk between the two. Wire it up as shown in Figure 6-19.

 BeagleBone Black runs at 3.3 V. When wiring other devices to it, ensure that they are also 3.3 V. The LaunchPad I'm using is 3.3 V, but many Arduinos are 5.0 V and thus won't work. Or worse, they might damage your Bone.

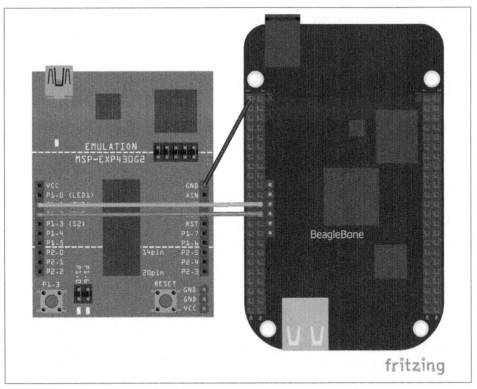

Figure 6-19. Wiring a LaunchPad to a Bone via the common serial port

Add the code (or *sketch*, as it's called in Arduino-speak) in Example 6-13 to a file called *launchPad.ino* and run it on your LaunchPad.

Example 6-13. LaunchPad code for communicating via the UART (launchPad.ino)

```
/*
  Tests connection to a BeagleBone
  Mark A. Yoder
  Waits for input on Serial Port
  g - Green toggle
  r - Red toggle
*/
char inChar = 0; // incoming serial byte
int red = 0;
int green = 0;

void setup()
{
  // initialize the digital pin as an output.
  pinMode(RED_LED, OUTPUT);           ❶
  pinMode(GREEN_LED, OUTPUT);
  // start serial port at 9600 bps:
  Serial.begin(9600);                 ❷
  Serial.print("Command (r, g): ");   ❸

  digitalWrite(GREEN_LED, green);     ❹
  digitalWrite(  RED_LED, red);
}

void loop()
{
  if(Serial.available() > 0 ) {       ❺
    inChar = Serial.read();
    switch(inChar) {                  ❻
      case 'g':
        green = ~green;
        digitalWrite(GREEN_LED, green);
        Serial.println("Green");
        break;
      case 'r':
        red = ~red;
        digitalWrite(RED_LED, red);
        Serial.println("Red");
        break;
    }
    Serial.print("Command (r, g): ");
  }
}
```

❶ Set the mode for the built-in red and green LEDs.

❷ Start the serial port at 9600 baud.

❸ Prompt the user, which in this case is the Bone.

❹ Set the LEDs to the current values of the red and green variables.

❺ Wait for characters to arrive on the serial port.

❻ After the characters are received, read it and respond to it.

On the Bone, add the script in Example 6-14 to a file called *launchPad.js* and run it.

Example 6-14. Code for communicating via the UART (launchPad.js)

```
#!/usr/bin/env node
// Need to add exports.serialParsers = m.module.parsers;
// to /usr/local/lib/node_modules/bonescript/serial.js
var b = require('bonescript');

var port = '/dev/tty01';                        ❶
var options = {
    baudrate: 9600,                             ❷
    parser: b.serialParsers.readline("\n")      ❸
};

b.serialOpen(port, options, onSerial);          ❹

function onSerial(x) {                           ❺
    console.log(x.event);
    if (x.err) {
        console.log('***ERROR*** ' + JSON.stringify(x));
    }
    if (x.event == 'open') {
        console.log('***OPENED***');
        setInterval(sendCommand, 1000);         ❻
    }
    if (x.event == 'data') {
        console.log(String(x.data));
    }
}

var command = ['r', 'g'];                        ❼
var commIdx = 1;

function sendCommand() {
    // console.log('Command: ' + command[commIdx]);
    b.serialWrite(port, command[commIdx++]);     ❽
    if(commIdx >= command.length) {              ❾
        commIdx = 0;
```

}

}

❶ Select which serial port to use. Figure 6-20 shows what's available. We've wired P9_24 and P9_26, so we are using serial port /dev/tty01. (Note that's the letter *O* and not the number *zero*.)

❷ Set the baudrate to 9600, which matches the setting on the LaunchPad.

❸ Read one line at a time up to the newline character (\n).

❹ Open the serial port and call onSerial() whenever there is data available.

❺ Determine what event has happened on the serial port and respond to it.

❻ If the serial port has been opened, start calling sendCommand() every 1000 ms.

❼ These are the two commands to send.

❽ Write the character out to the serial port and to the LaunchPad.

❾ Move to the next command.

P9				P8			
DGND	1	2	DGND	DGND	1	2	DGND
VDD_3V3	3	4	VDD_3V3	GPIO_38	3	4	GPIO_39
VDD_5V	5	6	VDD_5V	GPIO_34	5	6	GPIO_35
SYS_5V	7	8	SYS_5V	GPIO_66	7	8	GPIO_67
PWR_BUT	9	10	SYS_RESETN	GPIO_69	9	10	GPIO_68
UART4_RXD	11	12	GPIO_60	GPIO_45	11	12	GPIO_44
UART4_TXD	13	14	GPIO_50	GPIO_23	13	14	GPIO_26
GPIO_48	15	16	GPIO_51	GPIO_47	15	16	GPIO_46
GPIO_5	17	18	GPIO_4	GPIO_27	17	18	GPIO_65
UART1_RTSN	19	20	UART1_CTSN	GPIO_22	19	20	GPIO_63
UART2_TXD	21	22	UART2_RXD	GPIO_62	21	22	GPIO_37
GPIO_49	23	24	UART1_TXD	GPIO_36	23	24	GPIO_33
GPIO_117	25	26	UART1_RXD	GPIO_32	25	26	GPIO_61
GPIO_115	27	28	GPIO_113	GPIO_86	27	28	GPIO_88
GPIO_111	29	30	GPIO_112	GPIO_87	29	30	GPIO_89
GPIO_110	31	32	VDD_ADC	UART5_CTSN+	31	32	UART5_RTSN
AIN4	33	34	GNDA_ADC	UART4_RTSN	33	34	UART3_RTSN
AIN6	35	36	AIN5	UART4_CTSN	35	36	UART3_CTSN
AIN2	37	38	AIN3	UARR5_TXD+	37	38	UART5_RXD+
AIN0	39	40	AIN1	GPIO_76	39	40	GPIO_77
GPIO_20	41	42	UART3_TXD	GPIO_74	41	42	GPIO_75
DGND	43	44	DGND	GPIO_72	43	44	GPIO_73
DGND	45	46	DGND	GPIO_70	45	46	GPIO_71

Figure 6-20. Table of UART outputs

Discussion

When you run the script in Example 6-14, the Bone opens up the serial port and every second sends a new command, either r or g. The LaunchPad waits for the command and, when it arrives, responds by toggling the corresponding LED.

## CHAPTER 7

The Kernel

7.0 Introduction

The kernel is the heart of the Linux operating system. It's the software that takes the low-level requests, such as reading or writing files, or reading and writing general-purpose input/output (GPIO) pins, and maps them to the hardware. When you install a new version of the OS (Recipe 1.3), you get a certain version of the kernel.

You usually won't need to mess with the kernel, but sometimes you might want to try something new that requires a different kernel. This chapter shows how to switch kernels. The nice thing is you can have multiple kernels on your system at the same time and select from among them which to boot up.

 We assume here that you are logged on to your Bone as root and superuser privileges. You also need to be logged in to your Linux host computer as a nonsuperuser.

7.1 Updating the Kernel

Problem

You have an out-of-date kernel and want to want to make it current.

Solution

Use the following command to determine which kernel you are running:

```
bone# uname -a
Linux beaglebone 3.8.13-bone67 #1 SMP Wed Sep 24 21:30:03 UTC 2014 armv7l
GNU/Linux
```

The `3.8.13-bone67` string is the kernel version.

To update to the current kernel, ensure that your Bone is on the Internet (Recipe 5.13 or Recipe 5.11) and then run the following commands:

```
bone# apt-cache pkgnames | grep linux-image | sort | less
...
linux-image-3.15.8-armv7-x5
linux-image-3.15.8-bone5
linux-image-3.15.8-bone6
...
linux-image-3.16.0-rc7-bone1
...
linux-image-3.8.13-bone60
linux-image-3.8.13-bone61
linux-image-3.8.13-bone62
bone# apt-get install linux-image-3.14.23-ti-r35
bone# reboot
bone# uname -a
Linux beaglebone 3.14.23-ti-r35 #1 SMP PREEMPT Wed Nov 19 21:11:08 UTC 2014 armv7l
GNU/Linux
```

The first command lists the versions of the kernel that are available. The second command installs one. After you have rebooted, the new kernel will be running.

If the current kernel is doing its job adequately, you probably don't need to update, but sometimes a new software package requires a more up-to-date kernel. Fortunately, precompiled kernels are available and ready to download.

Discussion

If you have loaded multiple kernels, it's easy to switch between them without downloading them again:

```
bone# ls /boot
config-3.15.6-bone5            System.map-3.15.6-bone5
config-3.16.0-rc5-bone0       System.map-3.16.0-rc5-bone0
config-3.8.13-bone59          System.map-3.8.13-bone59
config-3.8.13-bone60          System.map-3.8.13-bone60
dtbs                          uboot
initrd.img-3.15.6-bone5       uEnv.txt
initrd.img-3.16.0-rc5-bone0   vmlinuz-3.15.6-bone5
initrd.img-3.8.13-bone59      vmlinuz-3.16.0-rc5-bone0
initrd.img-3.8.13-bone60      vmlinuz-3.8.13-bone59
SOC.sh                        vmlinuz-3.8.13-bone60
bone# head -4 /boot/uEnv.txt
#Docs: http://elinux.org/Beagleboard:U-boot_partitioning_layout_2.0
```

```
uname_r=3.8.13-bone67
```

The first command lists which kernels you have downloaded. The second command shows the first few lines of */boot/uEnv.txt*, which is the file that specifies which kernel to boot. Edit */boot/uEnv.txt* by changing `3.8.13-bone60` to the kernel you want and then reboot.

If there is a problem and you can't access the Bone through the network, see Recipe 5.5 for how to connect to the Bone by using an FTDI cable.

7.2 Building and Installing Kernel Modules

Problem

You need to use a peripheral for which there currently is no driver, or you need to improve the performance of an interface previously handled in user space.

Solution

The solution is to run in kernel space by building a kernel module. There are entire books on writing Linux Device Drivers (*http://bit.ly/1Fb0usf*). This recipe assumes that the driver has already been written and shows how to compile and install it. After you've followed the steps for this simple module, you will be able to apply them to any other module.

For our example module, add the code in Example 7-1 to a file called *hello.c*.

Example 7-1. Simple Kernel Module (hello.c)

```
#include <linux/module.h>      /* Needed by all modules */
#include <linux/kernel.h>      /* Needed for KERN_INFO */
#include <linux/init.h>        /* Needed for the macros */

static int __init hello_start(void)
{
    printk(KERN_INFO "Loading hello module...\n");
    printk(KERN_INFO "Hello, World!\n");
    return 0;
}

static void __exit hello_end(void)
{
    printk(KERN_INFO "Goodbye Boris\n");
}

module_init(hello_start);
module_exit(hello_end);
```

```
MODULE_AUTHOR("Boris Houndleroy");
MODULE_DESCRIPTION("Hello World Example");
MODULE_LICENSE("GPL");
```

When compiling on the Bone, all you need to do is load the Kernel Headers for the version of the kernel you're running:

```
bone# apt-get install linux-headers-`uname -r`
```

 The quotes around uname -r are backtick characters. On a United States keyboard, the backtick key is to the left of the 1 key.

This took a little more than three minutes on my Bone. The uname -r part of the command looks up what version of the kernel you are running and loads the headers for it.

Next, add the code in Example 7-2 to a file called *Makefile*.

Example 7-2. Simple Kernel Module (Makefile)

```
obj-m := hello.o
KDIR  := /lib/modules/$(shell uname -r)/build

all:
<TAB>make -C $(KDIR) M=$$PWD

clean:
<TAB>rm hello.mod.c hello.o modules.order hello.mod.o Module.symvers
```

 Replace the two instances of <TAB> with a tab character (the key left of the Q key on a United States keyboard). The tab characters are very important to makefiles and must appear as shown.

Now, compile the kernel module by using the make command:

```
bone# make
make -C /lib/modules/3.8.13-bone67/build \
     SUBDIRS=/root/cookbook-atlas/images/kernel/hello modules
make[1]: Entering directory `/usr/src/linux-headers-3.8.13-bone67'
  CC [M]  /root/cookbook-atlas/images/kernel/hello/hello.o
  Building modules, stage 2.
  MODPOST 1 modules
  CC      /root/cookbook-atlas/images/kernel/hello/hello.mod.o
  LD [M]  /root/cookbook-atlas/images/kernel/hello/hello.ko
```

```
make[1]: Leaving directory `/usr/src/linux-headers-3.8.13-bone67'
bone# ls
Makefile         hello.c    hello.mod.c  hello.o
Module.symvers   hello.ko   hello.mod.o  modules.order
```

Notice that several files have been created. *hello.ko* is the one you want. Try a couple of commands with it:

```
bone# modinfo hello.ko
filename:       /root/hello/hello.ko
srcversion:     87C6AEED7791B4B90C3B50C
depends:
vermagic:       3.8.13-bone67 SMP mod_unload modversions ARMv7 thumb2 p2v8
bone# insmod hello.ko
bone# dmesg | tail -4
[419313.320052] bone-iio-helper helper.15: ready
[419313.322776] bone-capemgr bone_capemgr.9: slot #8: Applied #1 overlays.
[491540.999431] Loading hello module...
[491540.999476] Hello world
```

The first command displays information about the module. The insmod command inserts the module into the running kernel. If all goes well, nothing is displayed, but the module does print something in the kernel log. The dmesg command displays the messages in the log, and the tail -4 command shows the last four messages. The last two messages are from the module. It worked!

Discussion

When you are finished with a module, you can also remove it:

```
bone# rmmod hello.ko
bone# dmesg | tail -4
[419313.322776] bone-capemgr bone_capemgr.9: slot #8: Applied #1 overlays.
[491540.999431] Loading hello module...
[491540.999476] Hello world
[492094.541102] Goodbye Mr.
```

The log message shows it has been successfully removed. You can also remove unneeded files:

```
bone# make clean
rm hello.mod.c hello.o modules.order hello.mod.o Module.symvers
bone# ls
Makefile  hello.c  hello.ko
```

7.3 Controlling LEDs by Using SYSFS Entries

Problem

You want to control the onboard LEDs from the command line.

Solution

On Linux, everything is a file (*http://bit.ly/1AjhWUW*); that is, you can access all the inputs and outputs, the LEDs, and so on by opening the right *file* and reading or writing to it. For example, try the following:

```
bone# cd /sys/class/leds/
bone# ls
beaglebone:green:usr0  beaglebone:green:usr2
beaglebone:green:usr1  beaglebone:green:usr3
```

What you are seeing are four directories, one for each onboard LED. Now try this:

```
bone# cd beaglebone\:green\:usr0
bone# ls
brightness  device  max_brightness  power  subsystem  trigger  uevent
bone# cat trigger
none nand-disk mmc0 mmc1 timer oneshot [heartbeat]
      backlight gpio cpu0 default-on transient
```

The first command changes into the directory for LED usr0, which is the LED closest to the edge of the board. The [heartbeat] indicates that the default trigger (behavior) for the LED is to blink in the heartbeat pattern. Look at your LED. Is it blinking in a heartbeat pattern?

Then try the following:

```
bone# echo none > trigger
bone# cat trigger
[none] nand-disk mmc0 mmc1 timer oneshot heartbeat
      backlight gpio cpu0 default-on transient
```

This instructs the LED to use none for a trigger. Look again. It should be no longer blinking.

Now, try turning it on and off:

```
bone# echo 1 > brightness
bone# echo 0 > brightness
```

The LED should be turning on and off with the commands.

Discussion

This recipe uses sysfs (*http://bit.ly/1C6OBUV*), a virtual file system provided by the Linux kernel to talk to hardware. There are many other things to explore in sysfs. Try looking around:

```
bone# cd /sys
bone# ls
block  bus  class  dev  devices  firmware  fs  kernel  module  power
bone# cd class
bone# ls
```

arvo	graphics	lcd	power_supply	scsi_host	usbmon
backlight	hidraw	leds	pps	sound	vc
bdi	hwmon	logibone	pwm	spidev	video4linux
block	i2c-adapter	mbox	pyra	spi_master	virtio-ports
bsg	i2c-dev	mdio_bus	rc	thermal	vtconsole
dma	input	mem	regulator	timed_output	watchdog
drm	isku	misc	rtc	tty	
dvb	kone	mmc_host	savu	ubi	
firmware	koneplus	mtd	scsi_device	udc	
gpio	kovaplus	net	scsi_disk	uio	

It's not hard to guess what's controlled by some of these files.

7.4 Controlling GPIOs by Using SYSFS Entries

Problem

You want to control a GPIO pin from the command line.

Solution

Recipe 7.3 introduces the sysfs. This recipe shows how to read and write a GPIO pin.

Reading a GPIO Pin via sysfs

Suppose that you want to read the state of the P9_42 GPIO pin. (Recipe 2.3 shows how to wire a switch to P9_42.) First, you need to map the P9 header location to GPIO number using Figure 7-1, which shows that P9_42 maps to GPIO 7.

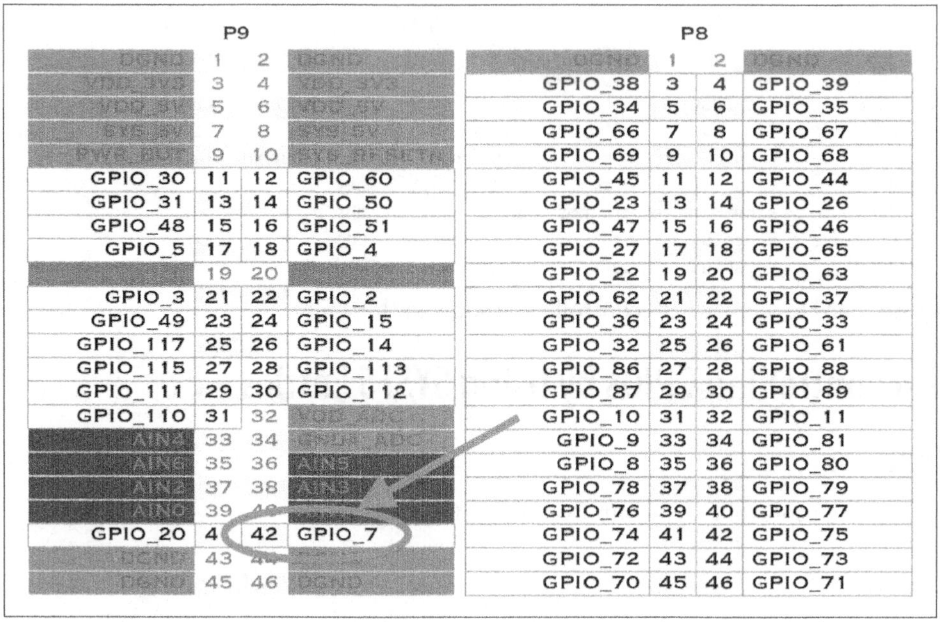

P9

	Pin	Pin	
DGND	1	2	DGND
VDD_3V3	3	4	VDD_3V3
VDD_5V	5	6	VDD_5V
SYS_5V	7	8	SYS_5V
PWR_BUT	9	10	SYS_RESETn
GPIO_30	11	12	GPIO_60
GPIO_31	13	14	GPIO_50
GPIO_48	15	16	GPIO_51
GPIO_5	17	18	GPIO_4
	19	20	
GPIO_3	21	22	GPIO_2
GPIO_49	23	24	GPIO_15
GPIO_117	25	26	GPIO_14
GPIO_115	27	28	GPIO_113
GPIO_111	29	30	GPIO_112
GPIO_110	31	32	VDD_ADC
AIN4	33	34	GNDA_ADC
AIN6	35	36	AIN5
AIN2	37	38	AIN3
AIN0	39	40	
GPIO_20	41	42	GPIO_7
DGND	43	44	
DGND	45	46	DGND

P8

	Pin	Pin	
DGND	1	2	DGND
GPIO_38	3	4	GPIO_39
GPIO_34	5	6	GPIO_35
GPIO_66	7	8	GPIO_67
GPIO_69	9	10	GPIO_68
GPIO_45	11	12	GPIO_44
GPIO_23	13	14	GPIO_26
GPIO_47	15	16	GPIO_46
GPIO_27	17	18	GPIO_65
GPIO_22	19	20	GPIO_63
GPIO_62	21	22	GPIO_37
GPIO_36	23	24	GPIO_33
GPIO_32	25	26	GPIO_61
GPIO_86	27	28	GPIO_88
GPIO_87	29	30	GPIO_89
GPIO_10	31	32	GPIO_11
GPIO_9	33	34	GPIO_81
GPIO_8	35	36	GPIO_80
GPIO_78	37	38	GPIO_79
GPIO_76	39	40	GPIO_77
GPIO_74	41	42	GPIO_75
GPIO_72	43	44	GPIO_73
GPIO_70	45	46	GPIO_71

Figure 7-1. Mapping P9_42 header position to GPIO 7

Next, change to the GPIO sysfs directory:

```
bone# cd /sys/class/gpio/
bone# ls
export  gpiochip0  gpiochip32  gpiochip64  gpiochip96  unexport
```

The ls command shows all the GPIO pins that have be exported. In this case, none have, so you see only the four GPIO controllers. Export using the export command:

```
bone# echo 7 > export
bone# ls
export  gpio7  gpiochip0  gpiochip32  gpiochip64  gpiochip96  unexport
```

Now you can see the *gpio7* directory. Change into the *gpio7* directory and look around:

```
bone# cd gpio7
bone# ls
active_low  direction  edge  power  subsystem  uevent  value
bone# cat direction
in
bone# cat value
0
```

Notice that the pin is already configured to be an input pin. (If it wasn't already configured that way, use echo in > direction to configure it.) You can also see that its current value is 0—that is, it isn't pressed. Try pressing and holding it and running again:

```
bone# cat value
1
```

The 1 informs you that the switch is pressed. When you are done with GPIO 7, you can always unexport it:

```
bone# cd ..
bone# echo 7 > unexport
bone# ls
export  gpiochip0  gpiochip32  gpiochip64  gpiochip96  unexport
```

Writing a GPIO Pin via sysfs

Now, suppose that you want to control an external LED. Recipe 3.2 shows how to wire an LED to P9_14. Figure 7-1 shows P9_14 is GPIO 50. Following the approach in Recipe 7.4, enable GPIO 50 and make it an output:

```
bone# cd /sys/class/gpio/
bone# echo 50 > export
bone# ls
gpio50  gpiochip0  gpiochip32  gpiochip64  gpiochip96
bone# cd gpio50
bone# ls
active_low  direction  edge  power  subsystem  uevent  value
bone# cat direction
in
```

By default, P9_14 is set as an input. Switch it to an output and turn it on:

```
bone# echo out > direction
bone# echo 1 > value
bone# echo 0 > value
```

The LED turns on when a 1 is written to value and turns off when a 0 is written.

Discussion

In Linux, interfaces to everything are turned into files. Although not everything is documented, you can learn more about the interfaces the kernel provides by checking out the ABI documentation (*http://bit.ly/1MtgLLp*). For example, the LED interface is documented in the ABI/testing/sysfs-class-led file (*http://bit.ly/1GEXGVS*). This documentation is part of the Linux kernel source tree, so you might find updates within the source tree of your particular kernel version.

The source tree for the kernel used in this book is located at *https://github.com/beagle board/linux/tree/3.8.*

7.5 Compiling the Kernel

Problem

You need to download, patch, and compile the kernel from its source code.

Solution

This is easier than it sounds, thanks to some very powerful scripts.

 Be sure to run this recipe on your host computer. The Bone has enough computational power to compile a module or two, but compiling the entire kernel takes lots of time and resourses.

Downloading and Compiling the Kernel

To download and compile the kernel, follow these steps:

```
host$ git clone https://github.com/RobertCNelson/bb-kernel.git ❶
host$ cd bb-kernel
host$ git tag ❷
host$ git checkout 3.8.13-bone60 -b v3.8.13-bone60 ❸
host$ ./build_kernel.sh ❹
```

❶ The first command clones a repository with the tools to build the kernel for the Bone.

❷ This command lists all the different versions of the kernel that you can build. You'll need to pick one of these. How do you know which one to pick? A good first step is to choose the one you are currently running. uname -a will reveal which one that is. When you are able to reproduce the current kernel, go to Linux Kernel Newbies (*http://kernelnewbies.org/*) to see what features are available in other kernels. LinuxChanges (*http://bit.ly/1AjiL00*) shows the features in the newest kernel and LinuxVersions (*http://bit.ly/1MrIHx3*) links to features of pervious kernels.

❸ When you know which kernel to try, use git checkout to check it out. This command checks out at tag 3.8.13-bone60 and creates a new branch, v3.8.13-bone60.

❹ build_kernel is the master builder. If needed, it will download the cross compilers needed to compile the kernel (linaro [*http://www.linaro.org/*] is the current cross compiler). If there is a kernel at *~/linux-dev*, it will use it; otherwise, it will download a copy to *bb-kernel/ignore/linux-src*. It will then patch the kernel so that it will run on the Bone.

After the kernel is patched, you'll see a screen similar to Figure 7-2, on which you can configure the kernel.

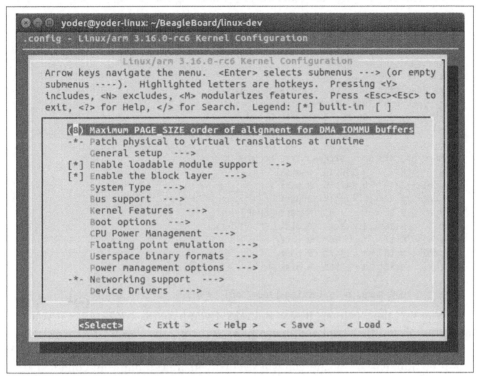

Figure 7-2. Kernel configuration menu

You can use the arrow keys to navigate. No changes need to be made, so you can just press the right arrow and Enter to start the kernel compiling. The entire process took about 25 minutes on my 8-core host.

The *bb-kernel/KERNEL* directory contains the source code for the kernel. The *bb-kernel/deploy* directory contains the compiled kernel and the files needed to run it.

Installing the Kernel on the Bone

To copy the new kernel and all its files to the microSD card, you need to halt the Bone, and then pull the microSD card out and put it in an microSD card reader on your host computer. Run Disk (see Recipe 1.3) to learn where the microSD card appears on your host (mine appears in */dev/sdb*). Then open the *bb-kernel/system.sh* file and find this line near the end:

```
#MMC=/dev/sde
```

Change that line to look like this (where /dev/sdb is the path to your device):

```
MMC=/dev/sdb
```

Now, while in the *bb-kernel* directory, run the following command:

```
host$ tools/install_kernel.sh
[sudo] password for yoder:

I see...
fdisk -l:
Disk /dev/sda: 160.0 GB, 160041885696 bytes
Disk /dev/sdb: 3951 MB, 3951034368 bytes
Disk /dev/sdc: 100 MB, 100663296 bytes

lsblk:
NAME    MAJ:MIN RM   SIZE RO TYPE MOUNTPOINT
sda     8:0      0 149.1G  0 disk
├─sda1  8:1      0 141.1G  0 part /
├─sda2  8:2      0    1K   0 part
└─sda5  8:5      0    8G   0 part [SWAP]
sdb     8:16     1  3.7G   0 disk
├─sdb1  8:17     1   16M   0 part
└─sdb2  8:18     1  3.7G   0 part
sdc     8:32     1   96M   0 disk
----------------------------
Are you 100% sure, on selecting [/dev/sdb] (y/n)? y
```

The script lists the partitions it sees and asks if you have the correct one. If you are sure, press Y, and the script will uncompress and copy the files to the correct locations on your card. When this is finished, eject your card, plug it into the Bone, and boot it up. Run uname -a, and you will see that you are running your compiled kernel.

Discussion

It is also possible to package your kernel as a *.deb* package and use dpkg -i to install the kernel. This also makes it possible to test bleeding-edge kernels from the Beagle-Board.org build server (*http://builds.beagleboard.org*).

7.6 Using the Installed Cross Compiler

Problem

You have followed the instructions in Recipe 7.5 and want to use the cross compiler it has downloaded.

You can cross-compile without installing the entire kernel source by running the following:

```
host$ sudo apt-get install gcc-arm-linux-gnueabihf
```

Then skip down to "Setting Up Variables" on page 240.

Solution

Recipe 7.5 installs a cross compiler, but you need to set up a couple of things so that it can be found. Recipe 7.5 installed the kernel and other tools in a directory called *bb-kernel*. Run the following commands to find the path to the cross compiler:

```
host$ cd bb-kernel/dl
host$ ls
gcc-linaro-arm-linux-gnueabihf-4.7-2013.04-20130415_linux
gcc-linaro-arm-linux-gnueabihf-4.7-2013.04-20130415_linux.tar.xz
```

Here, the path to the cross compiler contains the version number of the compiler. Yours might be different from mine. cd into it:

```
host$ cd gcc-linaro-arm-linux-gnueabihf-4.7-2013.04-20130415_linux
host$ ls
20130415-gcc-linaro-arm-linux-gnueabihf   bin   libexec
arm-linux-gnueabihf                        lib   share
```

At this point, we are interested in what's in *bin*:

```
host$ cd bin
host$ ls
arm-linux-gnueabihf-addr2line       arm-linux-gnueabihf-gfortran
arm-linux-gnueabihf-ar              arm-linux-gnueabihf-gprof
arm-linux-gnueabihf-as              arm-linux-gnueabihf-ld
arm-linux-gnueabihf-c++             arm-linux-gnueabihf-ld.bfd
arm-linux-gnueabihf-c++filt         arm-linux-gnueabihf-ldd
arm-linux-gnueabihf-cpp             arm-linux-gnueabihf-ld.gold
arm-linux-gnueabihf-ct-ng.config    arm-linux-gnueabihf-nm
arm-linux-gnueabihf-elfedit         arm-linux-gnueabihf-objcopy
arm-linux-gnueabihf-g++             arm-linux-gnueabihf-objdump
arm-linux-gnueabihf-gcc             arm-linux-gnueabihf-pkg-config
arm-linux-gnueabihf-gcc-4.7.3       arm-linux-gnueabihf-pkg-config-real
arm-linux-gnueabihf-gcc-ar          arm-linux-gnueabihf-ranlib
arm-linux-gnueabihf-gcc-nm          arm-linux-gnueabihf-readelf
arm-linux-gnueabihf-gcc-ranlib      arm-linux-gnueabihf-size
arm-linux-gnueabihf-gcov            arm-linux-gnueabihf-strings
arm-linux-gnueabihf-gdb             arm-linux-gnueabihf-strip
```

What you see are all the cross-development tools. You need to add this directory to the $PATH the shell uses to find the commands it runs:

```
host$ pwd
/home/yoder/BeagleBoard/bb-kernel/dl/\
    gcc-linaro-arm-linux-gnueabihf-4.7-2013.04-20130415_linux/bin
```

```
host$ echo $PATH
    /usr/local/sbin:/usr/local/bin:/usr/sbin:/usr/bin:/sbin:/bin:\
    /usr/games:/usr/local/games
```

The first command displays the path to the directory where the cross-development tools are located. The second shows which directories are searched to find commands to be run. Currently, the cross-development tools are not in the $PATH. Let's add it:

```
host$ export PATH=`pwd`:$PATH
host$ echo $PATH
/home/yoder/BeagleBoard/bb-kernel/dl/\
    gcc-linaro-arm-linux-gnueabihf-4.7-2013.04-20130415_linux/bin:\
    /usr/local/sbin:/usr/local/bin:/usr/sbin:/usr/bin:/sbin:/bin:\
    /usr/games:/usr/local/games
```

 Those are backtick characters (left of the "1" key on your keyboard) around pwd.

The second line shows the $PATH now contains the directory with the cross-development tools.

Setting Up Variables

Now, set up a couple of variables to know which compiler you are using:

```
host$ export ARCH=arm
host$ export CROSS_COMPILE=arm-linux-gnueabihf-
```

These lines set up the standard environmental variables so that you can determine which cross-development tools to use. Test the cross compiler by adding Example 7-3 to a file named *helloWorld.c*.

Example 7-3. Simple helloWorld.c to test cross compiling (helloWorld.c)

```
#include <stdio.h>

int main(int argc, char **argv) {
  printf("Hello, World! \n");
}
```

You can then cross-compile by using the following commands:

```
host$ ${CROSS_COMPILE}gcc helloWorld.c
host$ file a.out
a.out: ELF 32-bit LSB executable, ARM, version 1 (SYSV),
  dynamically linked (uses shared libs), for GNU/Linux 2.6.31,
  BuildID[sha1]=0x10182364352b9f3cb15d1aa61395aeede11a52ad, not stripped
```

The `file` command shows that `a.out` was compiled for an ARM processor.

Discussion

You can copy and run `a.out` on the Bone with the following commands:

```
host$ scp a.out root@192.168.7.2:.
a.out                          100% 8422     8.2KB/s   00:00
host$ ssh root@192.168.7.2 ./a.out
Hello, World!
```

The first line copies `a.out` to the home directory of `root` on the Bone at *192.168.7.2*.
The second line runs `a.out` on the Bone.

7.7 Applying Patches

Problem

You have a patch file that you need to apply to the kernel.

Solution

Example 7-4 shows a patch file that you can use on the kernel.

Example 7-4. Simple kernel patch file (hello.patch)

```
From eaf4f7ea7d540bc8bb57283a8f68321ddb4401f4 Mon Sep 17 00:00:00 2001
From: Jason Kridner <jdk@ti.com>
Date: Tue, 12 Feb 2013 02:18:03 +0000
Subject: [PATCH] hello: example kernel modules


---
 hello/Makefile   |    7 +++++++
 hello/hello.c    |   18 ++++++++++++++++++
 2 files changed, 25 insertions(+), 0 deletions(-)
 create mode 100644 hello/Makefile
 create mode 100644 hello/hello.c

diff --git a/hello/Makefile b/hello/Makefile
new file mode 100644
index 0000000..4b23da7
--- /dev/null
+++ b/hello/Makefile
@@ -0,0 +1,7 @@
+obj-m := hello.o
+
+PWD    := $(shell pwd)
+KDIR   := ${PWD}/..
+
```

```
+default:
+        make -C $(KDIR) SUBDIRS=$(PWD) modules
diff --git a/hello/hello.c b/hello/hello.c
new file mode 100644
index 0000000..157d490
--- /dev/null
+++ b/hello/hello.c
@@ -0,0 +1,22 @@
+#include <linux/module.h>        /* Needed by all modules */
+#include <linux/kernel.h>        /* Needed for KERN_INFO */
+#include <linux/init.h>          /* Needed for the macros */
+
+static int __init hello_start(void)
+{
+    printk(KERN_INFO "Loading hello module...\n");
+    printk(KERN_INFO "Hello, World!\n");
+    return 0;
+}
+
+static void __exit hello_end(void)
+{
+    printk(KERN_INFO "Goodbye Boris\n");
+}
+
+module_init(hello_start);
+module_exit(hello_end);
+
+MODULE_AUTHOR("Boris Houndleroy");
+MODULE_DESCRIPTION("Hello World Example");
+MODULE_LICENSE("GPL");
```

Here's how to use it:

1. Install the kernel sources (Recipe 7.5).

2. Change to the kernel directory (cd bb-kernel/KERNEL).

3. Add Example 7-4 to a file named *hello.patch* in the *bb-kernel/KERNEL* directory.

4. Run the following commands:

```
host$ cd bb-kernel/KERNEL
host$ patch -p1 < hello.patch
patching file hello/Makefile
patching file hello/hello.c
```

The output of the patch command apprises you of what it's doing. Look in the *hello* directory to see what was created:

```
host$ cd hello
host$ ls
hello.c  Makefile
```

Discussion

Recipe 7.2 shows how to build and install a module, and Recipe 7.8 shows how to create your own patch file.

7.8 Creating Your Own Patch File

Problem

You made a few changes to the kernel, and you want to share them with your friends.

Solution

Create a patch file that contains just the changes you have made. Before making your changes, check out a new branch:

```
host$ cd bb-kernel/KERNEL
host$ git status
# On branch master
nothing to commit (working directory clean)
```

Good, so far no changes have been made. Now, create a new branch:

```
host$ git checkout -b hello1
host$ git status
# On branch hello1
nothing to commit (working directory clean)
```

You've created a new branch called *hello1* and checked it out. Now, make whatever changes to the kernel you want. I did some work with a simple character driver that we can use as an example:

```
host$ cd bb-kernel/KERNEL/drivers/char/
host$ git status
# On branch hello1
# Changes not staged for commit:
#   (use "git add file..." to update what will be committed)
#   (use "git checkout -- file..." to discard changes in working directory)
#
#       modified:   Kconfig
#       modified:   Makefile
#
# Untracked files:
#   (use "git add file..." to include in what will be committed)
#
#       examples/
no changes added to commit (use "git add" and/or "git commit -a")
```

Add the files that were created and commit them:

```
host$ git add Kconfig Makefile examples
host$ git status
# On branch hello1
# Changes to be committed:
#   (use "git reset HEAD file..." to unstage)
#
#       modified:   Kconfig
#       modified:   Makefile
#       new file:   examples/Makefile
#       new file:   examples/hello1.c
#
host$ git commit -m "Files for hello1 kernel module"
[hello1 99346d5] Files for hello1 kernel module
 4 files changed, 33 insertions(+)
 create mode 100644 drivers/char/examples/Makefile
 create mode 100644 drivers/char/examples/hello1.c
```

Finally, create the patch file:

```
host$ git format-patch master --stdout > hello1.patch
```

Discussion

master is the branch you were on before creating the hello1 branch. The git
format-patch command creates a *hello1.patch* file, which includes instructions on
what files to change, create, or remove, to move from the master branch to the
hello1 branch.

## CHAPTER 8

Real-Time I/O

8.0 Introduction

Sometimes, when BeagleBone Black interacts with the physical world, it needs to respond in a timely manner. For example, your robot has just detected that one of the driving motors needs to turn a bit faster. Systems that can respond quickly to a real event are known as *real-time* systems. There are two broad categories of real-time systems: soft and hard.

In a *soft real-time* system, the real-time requirements should be met *most* of the time, where *most* depends on the system. A video playback system is a good example. The goal might be to display 60 frames per second, but it doesn't matter much if you miss a frame now and then. In a 100 percent *hard real-time* system, you can never fail to respond in time. Think of an airbag deployment system on a car. You can't even be 50 ms late.

Systems running Linux generally can't do 100 percent hard real-time processing, because Linux gets in the way. However, the Bone has an ARM processor running Linux and two additional 32-bit programmable real-time units (PRUs [*http://bit.ly/ 1EzTPZv*]) available to do real-time processing. Although the PRUs can achieve 100 percent hard real-time, they take some effort to use.

This chapter shows several ways to do real-time input/output (I/O), starting with the effortless, yet slower BoneScript and moving up with increasing speed (and effort) to using the PRUs.

In this chapter, as in the others, we assume that you are logged in as root (as indicated by the bone# prompt). This gives you quick access to the general-purpose input/output (GPIO) ports without having to use sudo each time.

8.1 I/O with BoneScript

Problem

You want to read an input pin and write it to the output as quickly as possible with BoneScript.

Solution

Recipe 2.3 shows how to read a pushbutton switch and Recipe 3.2 controls an external LED. This recipe combines the two to read the switch and turn on the LED in response to it. To make this recipe, you will need:

- Breadboard and jumper wires (see "Prototyping Equipment" on page 316)
- Pushbutton switch (see "Miscellaneous" on page 318)
- 220 Ω resistor (see "Resistors" on page 316)
- LED (see "Opto-Electronics" on page 318)

Wire up the pushbutton and LED as shown in Figure 8-1.

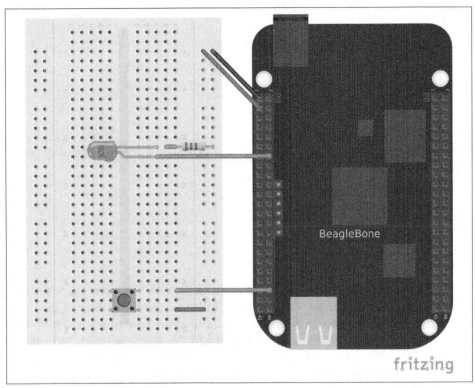

Figure 8-1. Diagram for wiring a pushbutton and LED with the LED attached to P9_12

The code in Example 8-1 reads GPIO port `P9_42`, which is attached to the pushbutton, and turns on the LED attached to `P9_12` when the button is pushed.

Example 8-1. Monitoring a pushbutton using a callback function (pushLED.js)

```
#!/usr/bin/env node
var b = require('bonescript');
var button = 'P9_42';
var LED     = 'P9_14';

b.pinMode(button, b.INPUT, 7, 'pulldown', 'fast', doAttach);
function doAttach(x) {
    if(x.err) {
        console.log('x.err = ' + x.err);
        return;
    }
    b.attachInterrupt(button, true, b.CHANGE, flashLED);
}

b.pinMode(LED,     b.OUTPUT);
```

```
function flashLED(x) {
    if(x.attached) {
        console.log("Interrupt handler attached");
        return;
    }
    console.log('x.value = ' + x.value);
    console.log('x.err   = ' + x.err);
    b.digitalWrite(LED, x.value);
}
```

Add the code to a file named *pushLED.js* and run it by using the following commands:

```
bone# chmod +x pushLED.js
bone# ./pushLED.js
```

Press ^C (Ctrl-C) to stop the code.

Discussion

BoneScript makes it easy to interact with the physical world. Here, every time the status of the pushbutton changes, `flashLED()` is called to update the state of the LED. BoneScript is more than fast enough to respond to a human pushing a button. But instead of the switch, what if you had a high-speed signal on the input, and what if you needed a device other than an LED for the output? Would BoneScript be fast enough? It depends on what you mean by "high speed." If BoneScript isn't fast enough, check out Recipe 8.2.

8.2 I/O with C and libsoc

Problem

You want to use the C language to process inputs in real time, or BoneScript isn't fast enough.

Solution

Recipe 8.1 shows how to control an LED with a pushbutton using BoneScript. This recipe accomplishes the same thing using C and `libsoc`. Recipe 5.21 shows how to use C and `libsoc` to access the GPIO pins.

Wire up the pushbutton and LED as shown in Figure 8-1. Follow the instructions in Recipe 5.21 to install `libsoc`. Then add the code in Example 8-2 to a file named *pushLED.c*.

Example 8-2. Code for reading a switch and blinking an LED using libsoc (pushLED.c)

```c
#include <stdio.h>
#include <stdlib.h>
#include <unistd.h>
#include <sys/wait.h>

#include "libsoc_gpio.h"
#include "libsoc_debug.h"

/**
 *
 * This gpio_test is intended to be run on BeagleBone
 * and reads pin P9_42 (gpio7) and write the value to P9_12 (gpio60).
 *
 * The GPIO_OUTPUT and INPUT defines can be changed to support any two pins.
 *
 */

#define GPIO_OUTPUT  60
#define GPIO_INPUT    7

// Create both gpio pointers
 gpio *gpio_output, *gpio_input;

static int interrupt_count = 0;

int callback_test(void* arg) {
  int* tmp_count = (int*) arg;
  int value;

  *tmp_count = *tmp_count + 1;
  value = libsoc_gpio_get_level (gpio_input);
  libsoc_gpio_set_level(gpio_output, value);
  // Comment out the following line to make the code respond faster
  printf("Got it (%d), button = %d\n", *tmp_count, value);

  return EXIT_SUCCESS;
}

int main(void) {
  // Enable debug output
  libsoc_set_debug(1);

  // Request gpios
  gpio_output = libsoc_gpio_request(GPIO_OUTPUT, LS_SHARED);
  gpio_input = libsoc_gpio_request(GPIO_INPUT, LS_SHARED);

  // Set direction to OUTPUT
  libsoc_gpio_set_direction(gpio_output, OUTPUT);

  // Set direction to INPUT
```

```
  libsoc_gpio_set_direction(gpio_input, INPUT);

  // Set edge to BOTH
  libsoc_gpio_set_edge(gpio_input, BOTH);

  // Set up callback
  libsoc_gpio_callback_interrupt(gpio_input, &callback_test,
                    (void*) &interrupt_count);

  printf("Push the button...\n");
  // Disaple debug output so the code will respond faster
  libsoc_set_debug(0);

  sleep(10);

  libsoc_set_debug(1);

  // Cancel the callback on interrupt
  libsoc_gpio_callback_interrupt_cancel(gpio_input);

  //If gpio_request was successful
  if (gpio_input) { // Free gpio request memory
    libsoc_gpio_free(gpio_input);
  }

  if (gpio_output) { // Free gpio request memory
    libsoc_gpio_free(gpio_output);
  }

  return EXIT_SUCCESS;
}
```

Compile and run the code:

```
bone# gcc -o pushLED pushLED.c /usr/local/lib/libsoc.so
bone# ./pushLED
libsoc-debug: debug enabled (libsoc_set_debug)
libsoc-gpio-debug: requested gpio (60, libsoc_gpio_request)
libsoc-gpio-debug: GPIO already exported (60, libsoc_gpio_request)
libsoc-gpio-debug: requested gpio (7, libsoc_gpio_request)
libsoc-gpio-debug: GPIO already exported (7, libsoc_gpio_request)
libsoc-gpio-debug: setting direction to out (60, libsoc_gpio_set_direction)
libsoc-gpio-debug: setting direction to in (7, libsoc_gpio_set_direction)
BOTH: 3
libsoc-gpio-debug: setting edge to both (7, libsoc_gpio_set_edge)
BOTH: 3
libsoc-gpio-debug: creating new callback (7, libsoc_gpio_callback_interrupt)
Push the button...
libsoc-debug: debug disabled (libsoc_set_debug)
Got it (1), button = 1
Got it (2), button = 0
Got it (3), button = 1
Got it (4), button = 1
```

```
Got it (5), button = 0
libsoc-debug: debug enabled (libsoc_set_debug)
libsoc-gpio-debug: callback thread was stopped
            (7, libsoc_gpio_callback_interrupt_cancel)
libsoc-gpio-debug: freeing gpio (7, libsoc_gpio_free)
libsoc-gpio-debug: freeing gpio (60, libsoc_gpio_free)
```

The code responds quickly to the pushbutton. If you need more speed, comment-out the `printf` command in the interrupt service routine.

Discussion

Because C code is compiled, it can respond more quickly and consistently to the interrupts than the BoneScript code. `libsoc` uses the Linux GPIO interface (Recipe 7.4) and is therefore portable to other Linux systems.

If you need even more speed, you can bypass the Linux interface and communicate directly with the hardware, as described in Recipe 8.3.

8.3 I/O with devmem2

Problem

Your C code using `libsoc` isn't responding fast enough to the input signal. You want to read the GPIO registers directly.

Solution

The solution is to use a simple utility called devmem2, with which you can read and write registers from the command line.

 This solution is much more involved than the previous ones. You need to understand binary and hex numbers and be able to read the AM335x Technical Reference Manual (*http://bit.ly/1B4Cm45*).

First, download and install devmem2:

```
bone# wget http://free-electrons.com/pub/mirror/devmem2.c
bone# gcc -o devmem2 devmem2.c
bone# mv devmem2 /usr/bin
```

This solution will read a pushbutton attached to P9_42 and flash an LED attached to P9_13. Note that this is a change from the previous solutions that makes the code used here much simpler. Wire up your Bone as shown in Figure 8-2.

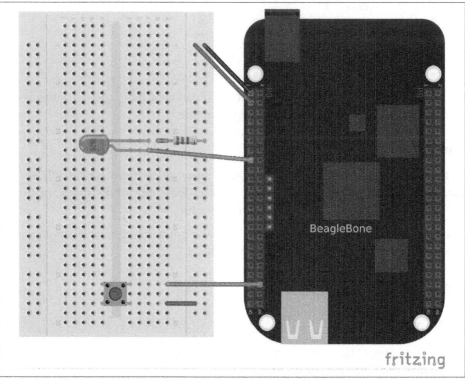

Figure 8-2. Diagram for wiring a pushbutton and LED with the LED attached to P9_13

Now, flash the LED attached to P9_13 using the Linux sysfs interface (Recipe 7.4). To do this, first look up which GPIO number P9_13 is attached to by referring to Figure 5-18. Finding P9_13 at GPIO 31, export GPIO 31 and make it an output:

```
bone# cd cd /sys/class/gpio/
bone# echo 31 > export
bone# cd gpio31
bone# echo out > direction
bone# echo 1 > value
bone# echo 0 > value
```

The LED will turn on when 1 is echoed into value and off when 0 is echoed.

Now that you know the LED is working, look up its memory address. This is where things get very detailed. First, download the AM335x Technical Reference Manual (*http://bit.ly/1B4Cm45*). Look up GPIO0 in the Memory Map chapter (sensors). Table 2-2 indicates that GPIO0 starts at address 0x44E0_7000. Then go to Section 25.4.1, "GPIO Registers." This shows that GPIO_DATAIN has an offset of 0x138, GPIO_CLEARDA TAOUT has an offset of 0x190, and GPIO_SETDATAOUT has an offset of 0x194.

This means you read from address 0x44E0_7000 + 0x138 = 0x44E0_7138 to see the status of the LED:

```
bone# devmem2 0x44E07138
/dev/mem opened.
Memory mapped at address 0xb6f8e000.
Value at address 0x44E07138 (0xb6f8e138): 0xC000C404
```

The returned value 0xC000C404 (1100 0000 0000 0000 1100 0100 0000 0100 in binary) has bit 31 set to 1, which means the LED is on. Turn the LED off by writing 0x80000000 (1000 0000 0000 0000 0000 0000 0000 0000 binary) to the GPIO_CLEARDATA register at 0x44E0_7000 + 0x190 = 0x44E0_7190:

```
bone# devmem2 0x44E07190 w 0x80000000
/dev/mem opened.
Memory mapped at address 0xb6fd7000.
Value at address 0x44E07190 (0xb6fd7190): 0x80000000
Written 0x80000000; readback 0x0
```

The LED is now off.

You read the pushbutton switch in a similar way. Figure 5-18 says P9_42 is GPIO 7, which means bit 7 is the state of P9_42. The devmem2 in this example reads 0x0, which means all bits are 0, including GPIO 7. Section 25.4.1 of the Technical Reference Manual instructs you to use offset 0x13C to read GPIO_DATAOUT. Push the pushbutton and run devmem2:

```
bone# devmem2 0x44e07138
/dev/mem opened.
Memory mapped at address 0xb6fe2000.
Value at address 0x44E07138 (0xb6fe2138): 0x4000C484
```

Here, bit 7 is set in 0x4000C484, showing the button is pushed.

Discussion

This is much more tedious than the previous methods, but it's what's necessary if you need to minimize the time to read an input. Recipe 8.4 shows how to read and write these addresses from C.

8.4 I/O with C and mmap()

Problem

Your C code using libsoc isn't responding fast enough to the input signal.

Solution

In smaller processors that aren't running an operating system, you can read and write a given memory address directly from C. With Linux running on Bone, many of the memory locations are hardware protected, so you can't accidentally access them directly.

This recipe shows how to use mmap() (memory map) to map the GPIO registers to an array in C. Then all you need to do is access the array to read and write the registers.

This solution is much more involved than the previous ones. You need to understand binary and hex numbers and be able to read the AM335x Technical Reference Manual.

This solution will read a pushbutton attached to P9_42 and flash an LED attached to P9_13. Note that this is a change from the previous solutions that makes the code used here much simpler.

See Recipe 8.3 for details on mapping the GPIO numbers to memory addresses.

Add the code in Example 8-3 to a file named *pushLEDmmap.h*.

Example 8-3. Memory address definitions (pushLEDmmap.h)

```
// From: http://stackoverflow.com/questions/13124271/driving-beaglebone-gpio
// -through-dev-mem
// user contributions licensed under cc by-sa 3.0 with attribution required
// http://creativecommons.org/licenses/by-sa/3.0/
// http://blog.stackoverflow.com/2009/06/attribution-required/
// Author: madscientist159 (http://stackoverflow.com/users/3000377/madscientist159)

#ifndef _BEAGLEBONE_GPIO_H_
#define _BEAGLEBONE_GPIO_H_

#define GPIO0_START_ADDR 0x44e07000
#define GPIO0_END_ADDR   0x44e08000
#define GPIO0_SIZE (GPIO0_END_ADDR - GPIO0_START_ADDR)

#define GPIO1_START_ADDR 0x4804C000
#define GPIO1_END_ADDR   0x4804D000
#define GPIO1_SIZE (GPIO1_END_ADDR - GPIO1_START_ADDR)
```

```
#define GPIO2_START_ADDR 0x41A4C000
#define GPIO2_END_ADDR   0x41A4D000
#define GPIO2_SIZE (GPIO2_END_ADDR - GPIO2_START_ADDR)

#define GPIO3_START_ADDR 0x41A4E000
#define GPIO3_END_ADDR   0x41A4F000
#define GPIO3_SIZE (GPIO3_END_ADDR - GPIO3_START_ADDR)

#define GPIO_DATAIN 0x138
#define GPIO_SETDATAOUT 0x194
#define GPIO_CLEARDATAOUT 0x190

#define GPIO_03  (1<<3)
#define GPIO_07  (1<<7)
#define GPIO_31  (1<<31)
#define GPIO_60  (1<<28)
#endif
```

Add the code in Example 8-4 to a file named *pushLEDmmap.c*.

Example 8-4. Code for directly reading memory addresses (pushLEDmmap.c)

```
// From: http://stackoverflow.com/questions/13124271/driving-beaglebone-gpio
// -through-dev-mem
// user contributions licensed under cc by-sa 3.0 with attribution required
// http://creativecommons.org/licenses/by-sa/3.0/
// http://blog.stackoverflow.com/2009/06/attribution-required/
// Author: madscientist159 (http://stackoverflow.com/users/3000377/madscientist159)
//
// Read one gpio pin and write it out to another using mmap.
// Be sure to set -O3 when compiling.
#include <stdio.h>
#include <stdlib.h>
#include <sys/mman.h>
#include <fcntl.h>
#include <signal.h>     // Defines signal-handling functions (i.e. trap Ctrl-C)
#include "pushLEDmmap.h"

// Global variables
int keepgoing = 1;      // Set to 0 when Ctrl-c is pressed

// Callback called when SIGINT is sent to the process (Ctrl-C)
void signal_handler(int sig) {
    printf( "\nCtrl-C pressed, cleaning up and exiting...\n" );
        keepgoing = 0;
}

int main(int argc, char *argv[]) {
    volatile void *gpio_addr;
    volatile unsigned int *gpio_datain;
    volatile unsigned int *gpio_setdataout_addr;
    volatile unsigned int *gpio_cleardataout_addr;
```

```
// Set the signal callback for Ctrl-C
signal(SIGINT, signal_handler);

int fd = open("/dev/mem", O_RDWR);

printf("Mapping %X - %X (size: %X)\n", GPIO0_START_ADDR, GPIO0_END_ADDR,
                                        GPIO0_SIZE);

gpio_addr = mmap(0, GPIO0_SIZE, PROT_READ | PROT_WRITE, MAP_SHARED, fd,
                 GPIO0_START_ADDR);

gpio_datain            = gpio_addr + GPIO_DATAIN;
gpio_setdataout_addr   = gpio_addr + GPIO_SETDATAOUT;
gpio_cleardataout_addr = gpio_addr + GPIO_CLEARDATAOUT;

if(gpio_addr == MAP_FAILED) {
    printf("Unable to map GPIO\n");
    exit(1);
}
printf("GPIO mapped to %p\n", gpio_addr);
printf("GPIO SETDATAOUTADDR mapped to %p\n", gpio_setdataout_addr);
printf("GPIO CLEARDATAOUT mapped to %p\n", gpio_cleardataout_addr);

printf("Start copying GPIO_07 to GPIO_31\n");
while(keepgoing) {
 if(*gpio_datain & GPIO_07) {
        *gpio_setdataout_addr= GPIO_31;
 } else {
        *gpio_cleardataout_addr = GPIO_31;
 }
    //usleep(1);
}

munmap((void *)gpio_addr, GPIO0_SIZE);
close(fd);
return 0;
}
```

Now, compile and run the code:

```
bone# gcc -O3 pushLEDmmap.c -o pushLEDmmap
bone# ./pushLEDmmap
Mapping 44E07000 - 44E08000 (size: 1000)
GPIO mapped to 0xb6fac000
GPIO SETDATAOUTADDR mapped to 0xb6fac194
GPIO CLEARDATAOUT mapped to 0xb6fac190
Start copying GPIO_07 to GPIO_31
^C
Ctrl-C pressed, cleaning up and exiting...
```

The code is in a tight `while` loop that checks the status of GPIO 7 and copies it to GPIO 31.

Discussion

This code is very fast at copying one GPIO bit to another. The disadvantage is that it uses 100 percent of the CPU. If you need the CPU to do other things, try uncommenting the `usleep()` command and recompiling. In my case, the response is fast, but only 30 percent of the CPU is used.

In this code, the `mmap()` command is used to map the GPIO0 registers to `gpio_addr`. The next few lines add the register offsets to `gpio_addr` to get the address for the registers being used. The code then goes into a tight `while` loop checking GPIO 7 and writing to GPIO 31.

The code uses `-O3` when compiling to tell the compiler to optimize the code.

This is likely the fastest way to process GPIOs with this kernel. The problem is that other programs are running on the Bone while this one is running. Sometimes, one of those programs can be using the CPU when the GPIO changes, and your program won't get control until several milliseconds later. Recipe 8.5 shows how to reconfigure the kernel to minimize the time your program can be interrupted.

8.5 Modifying the Linux Kernel to Use Xenomai

Problem

Your program can't get control of the processor quickly enough to do your real-time processing.

Solution

Xenomai (*http://xenomai.org/*) is a set of kernel patches that allow the Bone to respond to real-time events in a predicable way. These instructions are based on Bruno Martins' Xenomai on the BeagleBone Black in 14 easy steps (*http://bit.ly/1EXIEZu*).

The Xenomai kernel is used to provide real-time motion control within the Machinekit distribution for the Bone. You might get a quick start by checking out the images provided on the Machinekit blog (*http://bit.ly/1NK71xS*).

The Xenomai kernel is now available in the BeagleBoard.org Debian package feeds, and you can install it by using `apt-get install xenomai-runtime`. Nevertheless, we find it useful to include the details of building the Xenomai kernel.

This recipe requires downloading and patching the kernel. This is advanced work.

First, download, compile, and install the kernel, following the instructions in Recipe 7.5. This will ensure that everything is in place and working before you start changing things.

Next, visit Xenomai's download page (*http://bit.ly/1C6QxwN*) to find the latest version of Xenomai (at the time of this writing, it's 2.6.4). Download it with the following command (substituting the URL for the latest version you find on Xenomai's download page):

```
host$ wget http://download.gna.org/xenomai/stable/latest/xenomai-2.6.4.tar.bz2
```

Check out the latest kernel:

```
host$ cd bb-kernel/KERNEL
host$ uname -a
host$ git tags | sort | less
```

Pick the tag that is close to your current version of the kernel. This command checks out the 3.8.13-bone67 version of the kernel and creates a branch named xenomai:

```
host$ git checkout 3.8.13-bone67 -b xenomai
```

Patching the kernel

Patch the kernel by using the following commands:

```
host$ cd bb-kernel/KERNEL
host$ patch -p1 < ../../xenomai-2.6.4/ksrc/arch/arm/patches/\
      beaglebone/ipipe-core-3.8.13-beaglebone-pre.patch
host$ patch -p1 < ../../xenomai-2.6.4/ksrc/arch/arm/patches/\
      ipipe-core-3.8.13-arm-4.patch
host$ patch -p1 < ../../xenomai-2.6.4/ksrc/arch/arm/patches/\
      beaglebone/ipipe-core-3.8.13-beaglebone-post.patch
```

These commands assume bb-kernel and xenomai-2.6.4 are at the same directory level.

Then get one more patch from Xenomai's website (*http://bit.ly/1Mti9gQ*). Add the code in Example 8-5 to a file named *thumb.patch*.

Example 8-5. One last patch (thumb.patch)

```
diff --git a/arch/arm/kernel/entry-armv.S b/arch/arm/kernel/entry-armv.S
index e2bc263..6f4d9f0 100644
--- a/arch/arm/kernel/entry-armv.S
+++ b/arch/arm/kernel/entry-armv.S
@@ -469,6 +469,7 @@ __irq_usr:
 kuser_cmpxchg_check
 irq_handler
#ifdef CONFIG_IPIPE
+THUMB( it      ne )
 bne    __ipipe_ret_to_user_irqs_disabled
#endif /* CONFIG_IPIPE */
 get_thread_info tsk
@@ -767,6 +768,7 @@ ENTRY(ret_from_exception)
 disable_irq
 bl     __ipipe_check_root
 cmp    r0, #1
+THUMB( it      ne )
 bne    __ipipe_ret_to_user_irqs_disabled  @ Fast exit path over non-root domains
#endif /* CONFIG_IPIPE */
 get_thread_info tsk
```

Run the following command to apply the patch:

```
host$ patch -p1 < thumb.patch
```

Prepare the kernel:

```
host$ cd ../../xenomai-2.6.4/scripts
host$ ./prepare-kernel.sh --arch=arm --linux=../../bb-kernel/KERNEL/
```

Now you're ready to compile the kernel with these new patches:

```
host$ cd ../../bb-kernel
host$ tools/rebuild.sh
```

When the configuration menu displays (as shown in Figure 8-3), make the following changes:

1. Select "CPU Power Management" → "CPU Frequency scaling, disable [ ] CPU Frequency scaling."

2. Select "Real-time sub-system" →" Drivers" → "Testing drivers, enable everything."

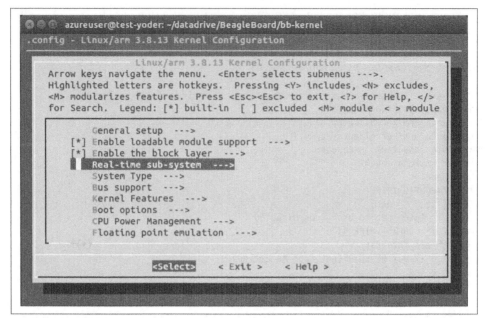

Figure 8-3. Configuing Xenomai

Then sit back and wait awhile. After the new kernel is built, install it following the instructions in "Installing the Kernel on the Bone" on page 237.

Setting up the Bone

You are now ready to test the new Xenomai kernel. Copy Xenomai to the Bone:

```
host$ scp -r xenomai-2.6.4 192.168.7.2:.
```

This command assumes that the Bone is at 192.168.7.2.

Run the following commands on the Bone:

```
bone# cd ~/xenomai-2.6.4
bone# ./configure CFLAGS="-march=armv7-a -mfpu=vfp3" LDFLAGS="-march=armv7-a \
        -mfpu=vfp3"
bone# make
bone# make install
```

`./configure` takes about a minute, `make` six more minutes, and `install` another minute.

Next, copy the drivers:

```
bone# cd; mkdir xenomai_drivers; cd xenomai_drivers

host$ cd bb-kernel/KERNEL/drivers/xenomai/testing
host$ scp *.ko 192.168.7.2:xenomai_drivers

bone# insmode xeno_klat.ko
```

Testing

Go ahead and run some tests:

```
bone# usr/xenomai/bin/latency
== Sampling period: 1000 us
== Test mode: periodic user-mode task
== All results in microseconds
warming up...
RTT|  00:00:01  (periodic user-mode task, 1000 us period, priority 99)
RTH|----lat min|----lat avg|----lat max|-overrun|---msw|---lat best|--lat worst
RTD|      3.624|      3.833|     20.458|       0|     0|      3.624|     20.458
RTD|      2.749|      3.749|     22.583|       0|     0|      2.749|     22.583
RTD|      2.749|      4.583|     25.499|       0|     0|      2.749|     25.499
RTD|      2.749|      3.708|     20.541|       0|     0|      2.749|     25.499
RTD|      2.749|      3.749|     21.916|       0|     0|      2.749|     25.499
---|-----------|-----------|-----------|--------|------|-------------------------
RTS|      2.749|      3.916|     25.499|       0|     0| 00:00:05/00:00:05
```

Discussion

The details of how to use Xenomai are beyond the scope of this book, but you can browse over to Xenomai's website (*http://xenomai.org/*) for examples and help on what to do next.

8.6 I/O with PRU Speak

Problem

You require better timing than running C on the ARM can give you.

Solution

The AM335x processor on the Bone has an ARM processor that is running Linux, but it also has two 32-bit PRUs that are available for processing I/O. It takes a fair amount of understanding to program the PRU. Fortunately, a 2014 Google Summer of Code (*http://bit.ly/1AjlarF*) project produced PRU Speak (*http://bit.ly/1wXT1eO*), an adaptation of the BotSpeak (*http://botspeak.org/*) language for the PRU. This solution shows how to use it.

Background

PRU Speak comprises three main parts:

Kernel

> The `pru_speak.ko` Linux kernel driver module loads PRU code and provides communication between the kernel and the firmware running on the PRU. An associated device tree overlay called `BB_PRUSPEAK-00A0.dtbo` configures the pin access and firmware loaded on to the PRUs.

Firmware

> `pru0_firmware` and `pru1_firmware` are the firmware files for each of the two PRUs.

Userspace

> A Python user space component includes `bs_tcp_server.py`; a TCP socket listener that provides an interface to the PRU Speak interpreter; and `pru_speak.py`, a library for invoking PRU Speak commands from within a Python script.

Prerequisites

Many of the pins easily accessible by the PRU are taken up by the HDMI interface discussed in Recipe 5.1. If you haven't already edited *uEnv.txt*, you can run these commands to disable the HDMI interface and reboot:

```
bone# sed -i \
 '/cape_disable=capemgr.disable_partno=BB-BONELT-HDMI,BB-BONELT-HDMIN$/ \
    \s/^#//' /boot/uEnv.txt
bone# shutdown -r now
```

To build the firmware files, you first need to install the TI PRU C compiler:

```
bone# apt-get update
bone# apt-get install ti-pru-cgt-installer
Reading package lists... Done
Building dependency tree
Reading state information... Done
The following NEW packages will be installed:
  ti-pru-cgt-installer
0 upgraded, 1 newly installed, 0 to remove and 30 not upgraded.
Need to get 13.1 kB of archives.
After this operation, 67.6 kB of additional disk space will be used.
Get:1 http://repos.rcn-ee.net/debian/ wheezy/main ti-pru-cgt-installer \
            all 2.1.0-1~bpo70+20141201+1 [13.1 kB]
Fetched 13.1 kB in 0s (38.1 kB/s)
Selecting previously unselected package ti-pru-cgt-installer.
(Reading database ... 59608 files and directories currently installed.)
Unpacking ti-pru-cgt-installer \
            (from .../ti-pru-cgt-installer_2.1.0-1~bpo70+20141201+1_all.deb) ...
Setting up ti-pru-cgt-installer (2.1.0-1~bpo70+20141201+1) ...
--2015-01-19 18:59:14--  http://downloads.ti.com/codegen/esd/cgt_public_sw/\
```

```
PRU/2.1.0/ti_cgt_pru_2.1.0_armlinuxa8hf_busybox_installer.sh
Resolving downloads.ti.com (downloads.ti.com)... 23.62.97.64, 23.62.97.66
Connecting to downloads.ti.com (downloads.ti.com)|23.62.97.64|:80... connected.
HTTP request sent, awaiting response... 200 OK
Length: 39666613 (38M) [application/x-sh]
Saving to: `ti_cgt_pru_2.1.0_armlinuxa8hf_busybox_installer.sh'

100%[====================================>] 39,666,613  75.3K/s   in 7m 39s

2015-01-19 19:06:54 (84.5 KB/s)-`ti_cgt_pru_2.1.0_arm..hf_busybox_installer.sh'\
         saved [39666613/39666613]

Installing PRU Code Generation tools version 2.1.0 into /
  please wait, or press CTRL-C to abort

Extracting archive
Installing files
[###################] 100%
Installed successfully into /
```

Installation

Clone the PRU Speak repository and run the *install.sh* shell script:

```
bone# git clone git://github.com/jadonk/pruspeak.git
bone# cd pruspeak
bone# ./install.sh
if [ ! -d "/lib/modules/3.8.13-bone68/build" ]; then apt-get install \
       linux-headers-3.8.13-bone68; fi
make -C /lib/modules/3.8.13-bone68/build M=/root/pruspeak/src/driver modules
make[1]: Entering directory `/usr/src/linux-headers-3.8.13-bone68'
  CC [M]  /root/pruspeak/src/driver/pru_speak.o
  Building modules, stage 2.
  MODPOST 1 modules
  CC      /root/pruspeak/src/driver/pru_speak.mod.o
  LD [M]  /root/pruspeak/src/driver/pru_speak.ko
make[1]: Leaving directory `/usr/src/linux-headers-3.8.13-bone68'
make -C /lib/modules/3.8.13-bone68/build M=/root/pruspeak/src/driver
modules_install
make[1]: Entering directory `/usr/src/linux-headers-3.8.13-bone68'
  INSTALL /root/pruspeak/src/driver/pru_speak.ko
  DEPMOD  3.8.13-bone68
make[1]: Leaving directory `/usr/src/linux-headers-3.8.13-bone68'
lnkpru -i/usr/share/ti/cgt-pru/lib pru0_firmware.obj pru0_syscall.obj \
       AM3359_PRU.cmd -o=pru0_firmware
lnkpru -i/usr/share/ti/cgt-pru/lib pru1_firmware.obj  AM3359_PRU.cmd
-o=pru1_firmware
install -m 444 pru0_firmware /lib/firmware
install -m 444 pru1_firmware /lib/firmware
dtc -O dtb -o BB-PRUSPEAK-00A0.dtbo -b 0 -@ BB-PRUSPEAK-00A0.dts
install -m 644 BB-PRUSPEAK-00A0.dtbo /lib/firmware
python setup.py build
running build
```

```
running build_py
python setup.py build
running build
running build_py
python setup.py install
running install
.
.
.
Processing pru_speak-0.0-py2.7.egg
Removing /usr/local/lib/python2.7/dist-packages/pru_speak-0.0-py2.7.egg
Copying pru_speak-0.0-py2.7.egg to /usr/local/lib/python2.7/dist-packages
pru-speak 0.0 is already the active version in easy-install.pth

Installed /usr/local/lib/python2.7/dist-packages/pru_speak-0.0-py2.7.egg
Processing dependencies for pru-speak==0.0
Searching for ply==3.4
Best match: ply 3.4
Processing ply-3.4-py2.7.egg
ply 3.4 is already the active version in easy-install.pth

Using /usr/local/lib/python2.7/dist-packages/ply-3.4-py2.7.egg
Finished processing dependencies for pru-speak==0.0
install -m 755 bs_tcp_server.py /usr/bin
install -m 755 bs_shell.py /usr/bin
```

Runing the Kernel Module

Use the modprobe command to load the kernel module and config-pin to load the
device tree overlay:

```
bone# modprobe pru_speak
bone# config-pin overlay BB-PRUSPEAK
```

The kernel module will need to be reloaded after rebooting, unless it's been added
to */etc/modules-load.d/modules.conf* and */boot/uEnv.txt*.

Running the Script

The PRU is now ready to run code. Add the code in Example 8-6 to a file named
pruSpeak.py. This code copies the input on P8_16 to the output on P9_31.

Example 8-6. Code for copying P8_16 to P9_31 using the PRU (pruSpeak.py)

```python
#!/usr/bin/env python
from pru_speak import pru_speak

# Copy the signal on pin P8_16 to P9_31
code = '''\
SCRIPT
SET var1, DIO[14]
SET DIO[0], var1
GOTO 0
ENDSCRIPT
RUN
'''

halt = '''\
HALT
'''

ret = pru_speak.execute_instruction(code)
print ret
```

Then run the code by using the following command:

```
bone# ./pruSpeak.py
Generating LALR tables
Initializing PRU Speak
0       0x110000FF      SET var1, DIO[14]
1       0x01400000      SET DIO[0], var1
2       0x15000000      GOTO 0
3       0x7F000000      ENDSCRIPT

[285212927, 20971520, 352321536, 2130706432]
[]
PRU Speak object deleted
```

The code will return immediately, but the PRU will continue to run.

The PRU has quick access to only a subset of GPIO pins. This example reads GPIO P8_16 and writes to P9_31, so you have to wire up the pushbutton and LED as shown in Figure 8-4.

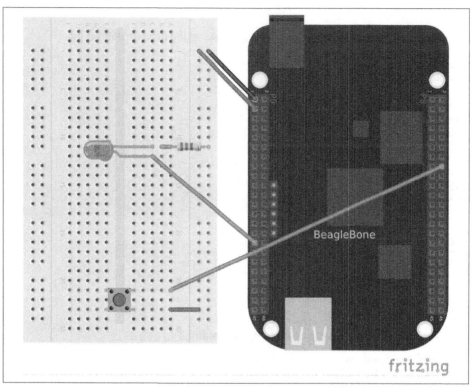

Figure 8-4. Diagram for wiring a pushbutton and LED for the PRU

Now, when you push the button, the LED should respond.

Discussion

In addition to invoking PRU Speak scripts via Python, it is also possible to open a shell or TCP socket to the interpreter. When running a script, it is also possible to modify variables to communicate with the running script.

Running PRU Speak via Shell

To invoke a shell interface to PRU Speak, run *bs_shell.py*:

```
bone# bs_shell.py
Generating LALR tables
Initializing PRU Speak
ps>SCRIPT
...SET DIO[0],var1
...GOTO 0
...ENDSCRIPT
...
0       0x01400001      SET DIO[0],var1
```

```
1       0x15000000      GOTO 0
2       0x7F000000      ENDSCRIPT

[]
ps>RUN
...
[20971521, 352321536, 2130706432]
[]
ps>SET var1,1
...
[1]
ps>SET var1,0
...
[0]
ps>Closing terminal
PRU Speak object deleted
```

Press Enter on a blank line to complete input to the shell. Press ^C (Ctrl-C) to exit the shell.

Running PRU Speak via TCP socket

Running the interpreter over a TCP socket enables tools like Labview to communicate with the Bone in a platform-independent way. Use netcat (nc) to interact with the interpreter from the command line:

```
bone# bs_tcp_server.py 2&>1 > /var/log/pruspeak.log &
[1] 2594
bone # nc localhost 6060
SCRIPT

SET DIO[0],1

WAIT 100

SET DIO[0],0

WAIT 100

GOTO 0

ENDSCRIPT

RUN

^C
```

Press ^C (Ctrl-C) to exit nc.

Currently, pressing ^D (Ctrl-D) within nc will cause the socket listener script to close.

You should now see the LED blinking.

Looking Ahead

At the time of this writing, PRU Speak is still under heavy development. Be sure to check the README file (*http://bit.ly/1B4DrsC*) for the latest installation and usage guidelines. Currently, only direct PRU GPIO access is implemented, but support for ADCs and additional GPIOs is in the works.

If you are looking to develop your own firmware for the PRUs, this example should provide a template for extending your own interface code.

## CHAPTER 9

Capes

9.0 Introduction

Previous chapters of this book show a variety of ways to interface BeagleBone Black to the physical world by using a breadboard and wiring to the P8 and P9 headers. This is a great approach because it's easy to modify your circuit to debug it or try new things. At some point, though, you might want a more permanent solution, either because you need to move the Bone and you don't want wires coming loose, or because you want to share your hardware with the masses.

You can easily expand the functionality of the Bone by adding a *cape* (*http://bit.ly/1wucweC*). A cape is simply a board—often a printed circuit board (PCB)—that connects to the P8 and P9 headers and follows a few standard pin usages. You can stack up to four capes onto the Bone. Capes can range in size from Bone-sized (Recipe 9.2) to much larger than the Bone (Recipe 9.1).

This chapter shows how to attach a couple of capes, move your design to a proto-board, then to a PCB, and finally on to mass production.

9.1 Using a Seven-Inch LCD Cape

Problem

You want to display the Bone's desktop on a portable LCD.

Solution

A number of LCD capes (*http://bit.ly/1AjlXJ9*) are built for the Bone, ranging in size from three to seven inches. This recipe attaches a seven-inch BeagleBone LCD7 (*http://bit.ly/1NK8Hra*) from CircuitCo (*http://circuitco.com/*) (shown in Figure 9-1) to the Bone.

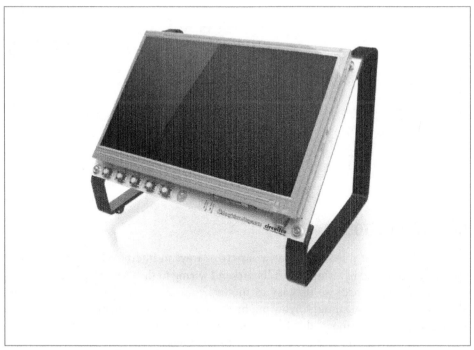

Figure 9-1. Seven-inch LCD from CircuitCo [1]

To make this recipe, you will need:

- Seven-inch LCD cape (see "Miscellaneous" on page 318)
- A 5 V power supply (see "Miscellaneous" on page 318)

Just attach the Bone to the back of the LCD, making sure pin 1 of P9 lines up with pin 1 of P9 on the LCD. Apply a 5 V power supply, and the desktop will appear on your LCD, as shown in Figure 9-2.

---

1 *Figure 9-1 was originally posted by CircuitCo at http://elinux.org/File:BeagleBone-LCD7-Front.jpg under a Creative Commons Attribution-ShareAlike 3.0 Unported License (http://creativecommons.org/licenses/by-sa/3.0/).*

Figure 9-2. Seven-inch LCD desktop

Attach a USB keyboard and mouse, and you have a portable Bone. Wireless keyboard and mouse combinations (*https://www.adafruit.com/products/922*) make a nice solution to avoid the need to add a USB hub.

Discussion

I have a *Rev A2* LCD display that was originally designed for the BeagleBone (White). I use it with the BeagleBone Black, but I have to remove the Bone from the LCD in order to update the onboard flash since the onboard flash signals are subjected to load and noises from the cape, creating problems with updating the onboard flash. Other than that, it works fine with BeagleBone Black.

9.2 Using a 128 x 128-Pixel LCD Cape

Problem

You want to use a small LCD to display things other than the desktop.

Solution

The MiniDisplay (*http://bit.ly/1xd0r8p*) is a 128 x 128 full-color LCD cape that just fits on the Bone, as shown in Figure 9-3.

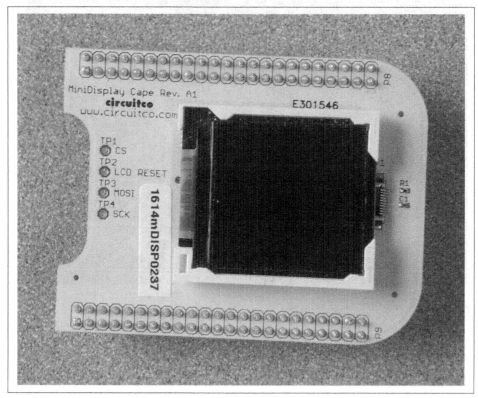

Figure 9-3. MiniDisplay 128 x 128-pixel LCD from CircuitCo

To make this recipe, you will need:

- MiniDisplay LCD cape (see "Miscellaneous" on page 318)

Attach to the Bone and apply power. Then run the following commands:

```
# From http://elinux.org/CircuitCo:MiniDisplay_Cape
# Datasheet:
# https://www.crystalfontz.com/products/document/3277/ST7735_V2.1_20100505.pdf
bone# wget http://elinux.org/images/e/e4/Minidisplay-example.tar.gz
bone# tar zmxvf Minidisplay-example.tar.gz
bone# cd minidisplay-example
bone# make
bone# ./minidisplay-test
Unable to initialize SPI: No such file or directory
Aborted
```

You might get a compiler warning, but the code should run fine.

The MiniDisplay uses the Serial Peripheral Interface (SPI) interface, and it's not initialized. The manufacturer's website (*http://bit.ly/1xd0r8p*) suggests enabling SPI0 by using the following commands:

```
bone# export SLOTS=/sys/devices/bone_capemgr.*/slots
bone# echo BB-SPIDEV0 > $SLOTS
```

Hmmm, something isn't working here. Here's how to see what happened:

```
bone# dmesg | tail
[  625.334497] bone_capemgr.9: part_number 'BB-SPIDEV0', version 'N/A'
[  625.334673] bone_capemgr.9: slot #11: generic override
[  625.334720] bone_capemgr.9: bone: Using override eeprom data at slot 11
[  625.334769] bone_capemgr.9: slot #11: 'Override Board Name,00A0,Override \
               Manuf,BB-SPIDEV0'
[  625.335026] bone_capemgr.9: slot #11: \Requesting part number/version based \
               'BB-SPIDEV0-00A0.dtbo
[  625.335076] bone_capemgr.9: slot #11: Requesting firmware \
               'BB-SPIDEV0-00A0.dtbo' \
               for board-name 'Override Board Name', version '00A0'
[  625.335144] bone_capemgr.9: slot #11: dtbo 'BB-SPIDEV0-00A0.dtbo' loaded; \
               converting to live tree
[  625.341842] bone_capemgr.9: slot #11: BB-SPIDEV0 conflict P9.21 \
               (#10:bspwm_P9_21_b) ❶
[  625.351296] bone_capemgr.9: slot #11: Failed verification
```

❶ Shows there is a conflict for pin P9_21: it's already configured for pulse width modulation (PWM).

Here's how to see what's already configured:

```
bone# cat $SLOTS
 0: 54:PF---
 1: 55:PF---
 2: 56:PF---
 3: 57:PF---
 4: ff:P-O-L Bone-LT-eMMC-2G,00A0,Texas Instrument,BB-BONE-EMMC-2G
 5: ff:P-O-L Bone-Black-HDMI,00A0,Texas Instrument,BB-BONELT-HDMI
 7: ff:P-O-L Override Board Name,00A0,Override Manuf,bspm_P9_42_27
 8: ff:P-O-L Override Board Name,00A0,Override Manuf,bspm_P9_41_27
 9: ff:P-O-L Override Board Name,00A0,Override Manuf,am33xx_pwm
10: ff:P-O-L Override Board Name,00A0,Override Manuf,bspwm_P9_21_b ❶
```

❶ You can see the eMMC, HDMI, and three PWMs are already using some of the pins. Slot 10 shows P9_21 is in use by a PWM.

You can unconfigure it by using the following commands:

```
bone# echo -10 > $SLOTS
bone# cat $SLOTS
 0: 54:PF---
 1: 55:PF---
 2: 56:PF---
 3: 57:PF---
 4: ff:P-O-L Bone-LT-eMMC-2G,00A0,Texas Instrument,BB-BONE-EMMC-2G
 5: ff:P-O-L Bone-Black-HDMI,00A0,Texas Instrument,BB-BONELT-HDMI
 7: ff:P-O-L Override Board Name,00A0,Override Manuf,bspm_P9_42_27
 8: ff:P-O-L Override Board Name,00A0,Override Manuf,bspm_P9_41_27
 9: ff:P-O-L Override Board Name,00A0,Override Manuf,am33xx_pwm
```

Now P9_21 is free for the MiniDisplay to use.

 In future Bone images, all of the pins will already be allocated as part of the main device tree using runtime pinmux helpers and configured at runtime using the config-pin utility (*http://bit.ly/ 1EXLeP2*). This would eliminate the need for device tree overlays in most cases.

Now, configure it for the MiniDisplay and run a test:

```
bone# echo BB-SPIDEV0 > $SLOTS
bone# ./minidisplay-test
```

You then see Boris, as shown in Figure 9-4.

Figure 9-4. MiniDisplay showing Boris [2]

---

2 *Figure 9-4 was originally posted by David Anders at http://elinux.org/File:Minidisplay-boris.jpg under a Creative Commons Attribution-ShareAlike 3.0 Unported License (http://creativecommons.org/licenses/by-sa/3.0/).*

Discussion

The MiniDisplay uses an SPI hardware interface, so it can run very fast. In this recipe, the data is sent to the display at 15 Mbps.

As an improvement to the MiniDisplay cape, you could add a cape EEPROM, which would avoid the need for users to manually load the device tree overlay. An additional improvement would be to include the support code in the default Bone Debian image by putting it in Debian package feed and including it in the image builder script (*http://bit.ly/1C6SaKM*). As a final improvement, you could provide a kernel driver to enable typical applications to display images on the screen or play videos.

9.3 Connecting Multiple Capes

Problem

You want to use more than one cape at a time.

Solution

First, look at each cape that you want to stack mechanically. Are they all using stacking headers like the ones shown in Figure 9-5? No more than one should be using non-stacking headers.

Figure 9-5. Stacking headers

Note that larger LCD panels might provide expansion headers, such as the ones shown in Figure 9-6, rather than the stacking headers, and that those can also be used for adding additional capes.

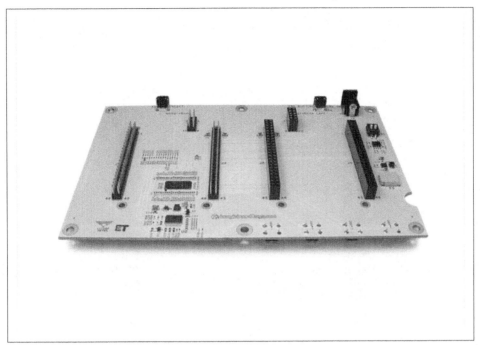

Figure 9-6. Back side of LCD7 cape [3]

Next, take a note of each pin utilized by each cape. The BeagleBone Capes catalog (*http://beaglebonecapes.com*) provides a graphical representation for the pin usage of most capes, as shown in Figure 9-7 for the Circuitco Audio Cape.

---

3 *Figure 9-6 was originally posted by CircuitCo at http://elinux.org/File:BeagleBone-LCD-Backside.jpg under a Creative Commons Attribution-ShareAlike 3.0 Unported License (http://creativecommons.org/licenses/by-sa/3.0/).*

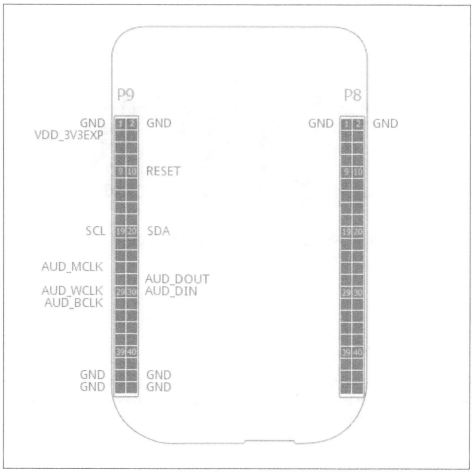

Figure 9-7. Pins utilized by CircuitCo Audio Cape [4]

In most cases, the same pin should never be used on two different capes, though in some cases, pins can be shared. Here are some exceptions:

GND
> The ground (GND) pins should be shared between the capes, and there's no need to worry about consumed resources on those pins.

---

4 *Figure 9-7 was originally posted by Djackson at http://elinux.org/File:Audio_pins_revb.png under a Creative Commons Attribution-ShareAlike 3.0 Unported License (http://creativecommons.org/licenses/by-sa/3.0/).*

VDD_3V3

The 3.3 V power supply (VDD_3V3) pins can be shared by all capes to supply power, but the total combined consumption of all the capes should be less than 500 mA (250 mA per VDD_3V3 pin).

VDD_5V

The 5.0 V power supply (VDD_5V) pins can be shared by all capes to supply power, but the total combined consumption of all the capes should be less than 2 A (1 A per VDD_5V pin). It is possible for one, and only one, of the capes to *provide* power to this pin rather than consume it, and it should provide at least 3 A to ensure proper system function. Note that when no voltage is applied to the DC connector, nor from a cape, these pins will not be powered, even if power is provided via USB.

SYS_5V

The regulated 5.0 V power supply (SYS_5V) pins can be shared by all capes to supply power, but the total combined consumption of all the capes should be less than 500 mA (250 mA per SYS_5V pin).

VADC *and* AGND

The ADC reference voltage pins can be shared by all capes.

I2C2_SCL *and* I2C2_SDA

I²C is a shared bus, and the I2C2_SCL and I2C2_SDA pins default to having this bus enabled for use by cape expansion ID EEPROMs.

Discussion

You have more issues to consider beyond just whether the pin functions are available, because every change to the topology of a circuit has at least *some* impact. There are changes to the transmission line behavior, changes to the capacitance and inductance of the signal, and more.

At low speeds, such as below 1 MHz, most of these effects can be ignored, but it is important to pay attention to the electrical specifications in the data sheet of any device you are using. When you try out your circuits, use an oscilloscope at various points on the significant signals to ensure that you see the behavior you expect. Don't assume that just because things seem to be generally working that there isn't something marginal about your construction. The manufacturer of your capes likely didn't think a lot about them being used with other capes.

9.4 Moving from a Breadboard to a Protoboard

Problem

You have your circuit working fine on the breadboard, but you want a more reliable solution.

Solution

Solder your components to a protoboard.

To make this recipe, you will need:

- Protoboard (see "Prototyping Equipment" on page 316)
- Soldering iron (see "Miscellaneous" on page 318)
- Your other components

Many places make premade circuit boards that are laid out like the breadboard we have been using. Figure 9-8 shows the BeagleBone Breadboard (*http://bit.ly/1HCwtB4*), which is just one protoboard option.

Figure 9-8. BeagleBone breadboard [5]

You just solder your parts on the protoboard as you had them on the breadboard.

---

[5] *Figure 9-8 was originally posted by William Traynor at http://elinux.org/File:BeagleBone-Breadboard.jpg under a Creative Commons Attribution-ShareAlike 3.0 Unported License (http://creativecommons.org/licenses/by-sa/3.0/).*

Discussion

You can find many tutorials online that show how to move from a breadboard to a protoboard, including tips on soldering. Evil Mad Scientist (*http://bit.ly/18AzcOW*) is one such site where you can find a good tutorial.

Many protocapes will mount on BeagleBone Black and provide a place to solder components. See the BeagleBone Capes catalog (*http://bit.ly/1AjlXJ9*) for a selection. For example, if you want something that more closely matches a larger and more typical breadboard layout, consider the Perma-Proto Breadboard PCB from Adafruit (*https://www.adafruit.com/products/590*).

Using a protoboard is a useful solution if you want your parts soldered down but don't want to go through the trouble of making a PCB.

9.5 Creating a Prototype Schematic

Problem

You've wired up a circuit on a breadboard. How do you turn that prototype into a schematic others can read and that you can import into other design tools?

Solution

In "About the Diagrams" on page 11, we introduced Fritzing as a useful tool for drawing block diagrams. Fritzing can also do circuit schematics and printed-circuit layout. For example, Figure 9-9 shows a block diagram for a simple robot controller (*quickBot.fzz* is the name of the Fritzing file used to create the diagram).

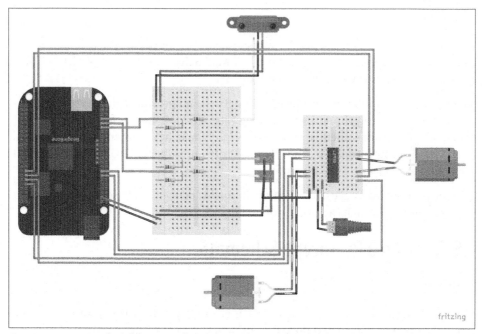

Figure 9-9. A simple robot controller diagram (quickBot.fzz)

The controller has an H-bridge to drive two DC motors (Recipe 4.3), an IR range sensor, and two headers for attaching analog encoders for the motors. Both the IR sensor and the encoders have analog outputs that exceed 1.8 V, so each is run through a voltage divider (two resistors) to scale the voltage to the correct range (see Recipe 2.6 for a voltage divider example).

Figure 9-10 shows the schematic automatically generated by Fritzing. It's a mess. It's up to you to fix it.

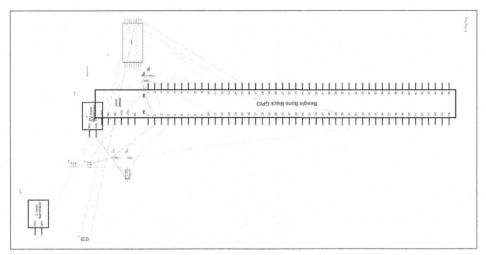

Figure 9-10. Automatically generated schematic

Figure 9-11 shows my cleaned-up schematic. I did it by moving the parts around until it looked better.

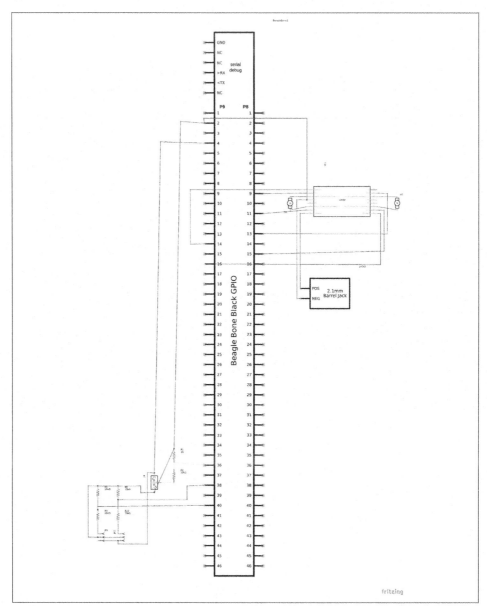

Figure 9-11. Cleaned-up schematic

Discussion

In Fritzing, you can zoom in on the schematic and check your wiring. In the zoomed-in section shown in Figure 9-12, you can see the pin labels on the L293D. You also can see I left a *V+* pin unattached. This is the sort of error that's easier to catch in the schematic view than the diagram view.

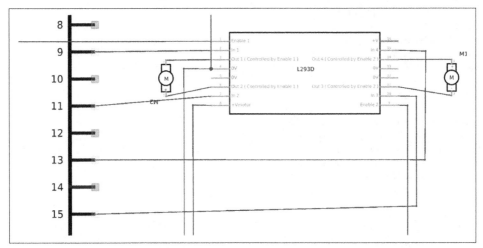

Figure 9-12. Zoomed-in schematic

You might find that you want to create your design in a more advanced design tool, perhaps because it has the library components you desire, it integrates better with other tools you are using, or it has some other feature (such as simulation) of which you'd like to take advantage.

9.6 Verifying Your Cape Design

Problem

You've got a design. How do you quickly verify that it works?

Solution

To make this recipe, you will need:

- An oscilloscope (see "Miscellaneous" on page 318)

Break down your design into functional subcomponents and write tests for each. Use components you already know are working, such as the onboard LEDs, to display the test status with the code in Example 9-1.

Example 9-1. Testing the quickBot motors interface (quickBot_motor_test.js)

```
#!/usr/bin/env node
var b = require('bonescript');
```

```
var M1_SPEED    = 'P9_16'; ❶
var M1_FORWARD  = 'P8_15';
var M1_BACKWARD = 'P8_13';
var M2_SPEED    = 'P9_14';
var M2_FORWARD  = 'P8_9';
var M2_BACKWARD = 'P8_11';
var freq = 50; ❷
var fast = 0.95;
var slow = 0.7;
var state = 0; ❸

b.pinMode(M1_FORWARD, b.OUTPUT); ❹
b.pinMode(M1_BACKWARD, b.OUTPUT);
b.pinMode(M2_FORWARD, b.OUTPUT);
b.pinMode(M2_BACKWARD, b.OUTPUT);
b.analogWrite(M1_SPEED, 0, freq); ❺
b.analogWrite(M2_SPEED, 0, freq);

updateMotors(); ❻

function updateMotors() { ❻
    //console.log("Setting state = " + state); ❼
    updateLEDs(state); ❼
    switch(state) { ❽
        case 0:
        default:
            M1_set(0); ❽
            M2_set(0);
            state = 1; ❸
            break;
        case 1:
            M1_set(slow);
            M2_set(slow);
            state = 2;
            break;
        case 2:
            M1_set(slow);
            M2_set(-slow);
            state = 3;
            break;
        case 3:
            M1_set(-slow);
            M2_set(slow);
            state = 4;
            break;
        case 4:
            M1_set(fast);
            M2_set(fast);
            state = 0;
            break;
    }
    setTimeout(updateMotors, 2000); ❸
```

```
}

function updateLEDs(state) { ❼
    switch(state) {
    case 0:
        b.digitalWrite("USR0", b.LOW);
        b.digitalWrite("USR1", b.LOW);
        b.digitalWrite("USR2", b.LOW);
        b.digitalWrite("USR3", b.LOW);
        break;
    case 1:
        b.digitalWrite("USR0", b.HIGH);
        b.digitalWrite("USR1", b.LOW);
        b.digitalWrite("USR2", b.LOW);
        b.digitalWrite("USR3", b.LOW);
        break;
    case 2:
        b.digitalWrite("USR0", b.LOW);
        b.digitalWrite("USR1", b.HIGH);
        b.digitalWrite("USR2", b.LOW);
        b.digitalWrite("USR3", b.LOW);
        break;
    case 3:
        b.digitalWrite("USR0", b.LOW);
        b.digitalWrite("USR1", b.LOW);
        b.digitalWrite("USR2", b.HIGH);
        b.digitalWrite("USR3", b.LOW);
        break;
    case 4:
        b.digitalWrite("USR0", b.LOW);
        b.digitalWrite("USR1", b.LOW);
        b.digitalWrite("USR2", b.LOW);
        b.digitalWrite("USR3", b.HIGH);
        break;
    }
}

function M1_set(speed) { ❽
    speed = (speed > 1) ? 1 : speed; ❾
    speed = (speed < -1) ? -1 : speed;
    b.digitalWrite(M1_FORWARD, b.LOW);
    b.digitalWrite(M1_BACKWARD, b.LOW);
    if(speed > 0) {
        b.digitalWrite(M1_FORWARD, b.HIGH);
    } else if(speed < 0) {
        b.digitalWrite(M1_BACKWARD, b.HIGH);
    }
    b.analogWrite(M1_SPEED, Math.abs(speed), freq); ❿
}

function M2_set(speed) {
    speed = (speed > 1) ? 1 : speed;
```

```
    speed = (speed < -1) ? -1 : speed;
    b.digitalWrite(M2_FORWARD, b.LOW);
    b.digitalWrite(M2_BACKWARD, b.LOW);
    if(speed > 0) {
        b.digitalWrite(M2_FORWARD, b.HIGH);
    } else if(speed < 0) {
        b.digitalWrite(M2_BACKWARD, b.HIGH);
    }
    b.analogWrite(M2_SPEED, Math.abs(speed), freq);
}
```

❶ Define each pin as a variable. This makes it easy to change to another pin if you decide that is necessary.

❷ Make other simple parameters variables. Again, this makes it easy to update them. When creating this test, I found that the PWM frequency to drive the motors needed to be relatively low to get over the kickback shown in Figure 9-13. I also found that I needed to get up to about 70 percent duty cycle for my circuit to reliably start the motors turning.

❸ Use a simple variable such as state to keep track of the test phase. This is used in a switch statement to jump to the code to configure for that test phase and updated after configuring for the current phase in order to select the next phase. Note that the next phase isn't entered until after a two-second delay, as specified in the call to setTimeout().

❹ Perform the initial setup of all the pins.

❺ The first time a PWM pin is used, it is configured with the update frequency. It is important to set this just once to the right frequency, because other PWM channels might use the same PWM controller, and attempts to reset the PWM frequency might fail. The pinMode() function doesn't have an argument for providing the update frequency, so use the analogWrite() function, instead. You can review using the PWM in Recipe 4.1.

❻ updateMotors() is the test function for the motors and is defined after all the setup and initialization code. The code calls this function every two seconds using the setTimeout() JavaScript function. The first call is used to prime the loop.

❼ The call to console.log() was initially here to observe the state transitions in the debug console, but it was replaced with the updateLEDs() call. Using the USER LEDs makes it possible to note the state transitions without having visibility of the debug console. updateLEDs() is defined later.

❽ The `M1_set()` and `M2_set()` functions are defined near the bottom and do the work of configuring the motor drivers into a particular state. They take a single argument of `speed`, as defined between `-1` (maximum reverse), `0` (stop), and `1` (maximum forward).

❾ Perform simple bounds checking to ensure that speed values are between `-1` and `1`.

❿ The `analogWrite()` call uses the absolute value of `speed`, making any negative numbers a positive magnitude.

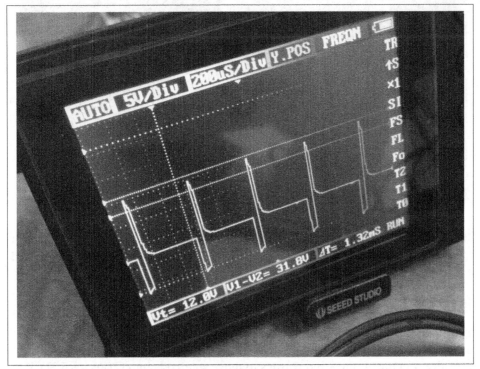

Figure 9-13. quickBot motor test showing kickback

Using the solution in Recipe 1.8, you can untether from your coding station to test your design at your lab workbench, as shown in Figure 9-14.

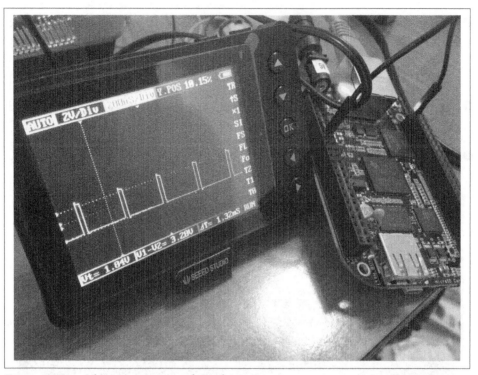

Figure 9-14. quickBot motor test code under scope

SparkFun provides a useful guide to using an oscilloscope (*http://bit.ly/18AzuoR*). You might want to check it out if you've never used an oscilloscope before. Looking at the stimulus you'll generate *before* you connect up your hardware will help you avoid surprises.

Discussion

As with most of the examples in this book, this recipe provides solutions for interfacing with various devices using JavaScript and the BoneScript library. For higher performance, you might often find reasons to move your code into C (Recipe 5.21), into the kernel (Recipe 7.2), or onto the PRUs (Recipe 8.6). Nevertheless, having something that quickly verifies the electrical and mechanical aspects of your design enables you to move forward with those aspects while you continue to refine your software.

9.7 Laying Out Your Cape PCB

Problem

You've generated a diagram and schematic for your circuit and verified that they are correct. How do you create a PCB?

Solution

If you've been using Fritzing, all you need to do is click the PCB tab, and there's your board. Well, almost. Much like the schematic view shown in Recipe 9.5, you need to do some layout work before it's actually usable. I just moved the components around until they seemed to be grouped logically and then clicked the Autoroute button. After a minute or two of trying various layouts, Fritzing picked the one it determined to be the best. Figure 9-15 shows the results.

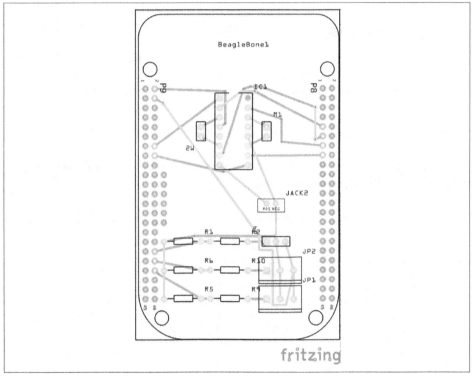

Figure 9-15. Simple robot PCB

The Fritzing pre-fab web page (*http://bit.ly/1HCxokQ*) has a few helpful hints, including checking the widths of all your traces and cleaning up any questionable routing created by the autorouter.

Discussion

The PCB in Figure 9-15 is a two-sided board. One color (or shade of gray in the printed book) represents traces on one side of the board, and the other color (or shade of gray) is the other side. Sometimes, you'll see a trace come to a small circle and then change colors. This is where it is switching sides of the board through what's called a *via*. One of the goals of PCB design is to minimize the number of vias.

Figure 9-15 wasn't my first try or my last. My approach was to see what was needed to hook where and move the components around to make it easier for the autorouter to carry out its job.

 There are entire books and websites dedicated to creating PCB layouts. Look around and see what you can find. SparkFun's guide to making PCBs (*http://bit.ly/1wXTLki*) is particularly useful.

Customizing the Board Outline

One challenge that slipped my first pass review was the board outline. The part we installed in "About the Diagrams" on page 11 is meant to represent BeagleBone Black, not a cape, so the outline doesn't have the notch cut out of it for the Ethernet connector.

The Fritzing custom PCB outline page (*http://bit.ly/1xd1aGV*) describes how to create and use a custom board outline. Although it is possible to use a drawing tool like Inkscape (*https://inkscape.org/en/*), I chose to use the SVG *path* command (*http://bit.ly/1b2aZmn*) directly to create Example 9-2.

Example 9-2. Outline SVG for BeagleBone cape (beaglebone_cape_boardoutline.svg)

```
<?xml version='1.0' encoding='UTF-8' standalone='no'?>
<svg xmlns="http://www.w3.org/2000/svg" version="1.1"
    width="306"  height="193.5"><!--❶-->
 <g id="board"><!--❷-->
  <path fill="#338040" id="boardoutline" d="M 22.5,0 l 0,56 L 72,56
     q 5,0 5,5 l 0,53.5 q 0,5 -5,5 L 0,119.5 L 0,171 Q 0,193.5 22.5,193.5
     l 238.5,0 c 24.85281,0 45,-20.14719 45,-45 L 306,45
     C 306,20.14719 285.85281,0 261,0 z"/><!--❸-->
 </g>
</svg>
```

❶ This is a standard SVG header. The width and height are set based on the Beagle-Bone outline provided in the Adafruit library.

❷ Fritzing requires the element to be within a layer called board.

❸ Fritzing requires the color to be #338040 and the layer to be called boardoutline. The units end up being 1/90 of an inch. That is, take the numbers in the SVG code and divide by 90 to get the numbers from the System Reference Manual.

The measurements are taken from the BeagleBone Black System Reference Manual (*http://bit.ly/1C5rSa8*), as shown in Figure 9-16.

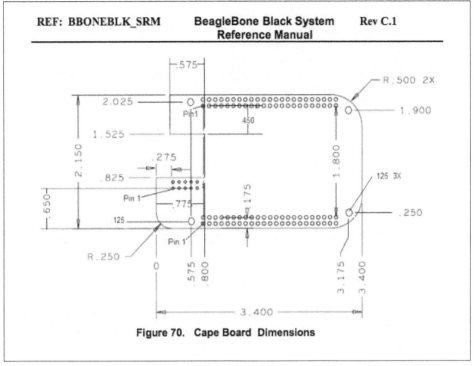

Figure 70. Cape Board Dimensions

Figure 9-16. Cape dimensions

You can observe the rendered output of Example 9-2 quickly by opening the file in a web browser, as shown in Figure 9-17.

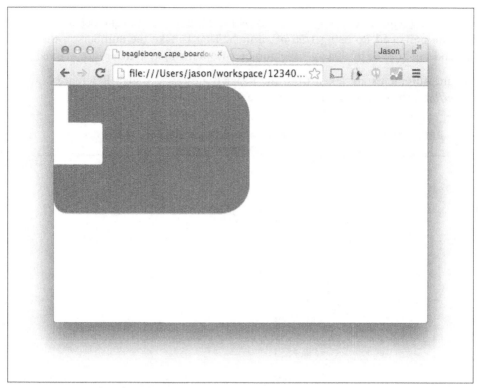

Figure 9-17. Rendered cape outline in Chrome

After you have the SVG outline, you'll need to select the PCB in Fritzing and select a custom shape in the Inspector box. Begin with the original background, as shown in Figure 9-18.

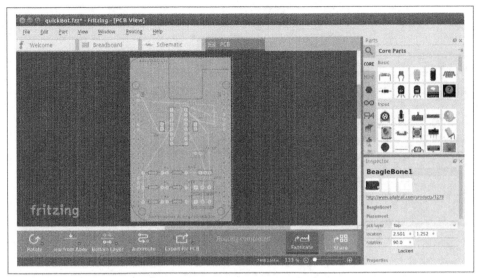

Figure 9-18. PCB with original board, without notch for Ethernet connector

Hide all but the Board Layer (Figure 9-19).

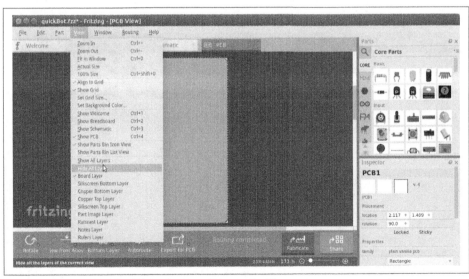

Figure 9-19. PCB with all but the Board Layer hidden

Select the PCB1 object and then, in the Inspector pane, scroll down to the "load image file" button (Figure 9-20).

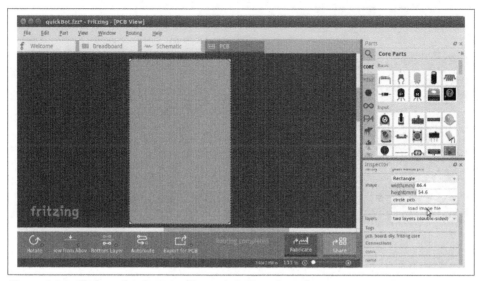

Figure 9-20. Clicking :load image file: with PCB1 selected

Navigate to the *beaglebone_cape_boardoutline.svg* file created in Example 9-2, as shown in Figure 9-21.

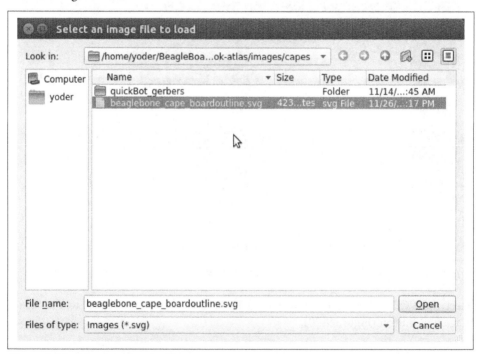

Figure 9-21. Selecting the .svg file

Turn on the other layers and line up the Board Layer with the rest of the PCB, as shown in Figure 9-22.

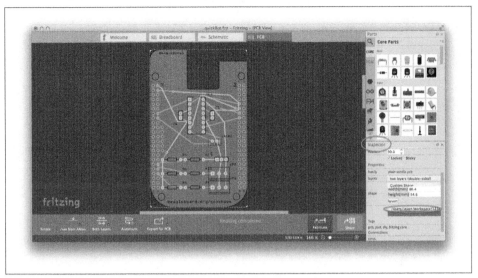

Figure 9-22. PCB Inspector

Now, you can save your file and send it off to be made, as described in Recipe 9.9.

PCB Design Alternatives

There are other free PCB design programs. Here are a few.

EAGLE

Eagle PCB (*http://www.cadsoftusa.com/*) and DesignSpark PCB (*http://bit.ly/ 19cbwS0*) are two popular design programs. Many capes (and other PCBs) are designed with Eagle PCB, and the files are available. For example, the MiniDisplay cape (Recipe 9.2) has the schematic shown in Figure 9-23 and PCB shown in Figure 9-24.

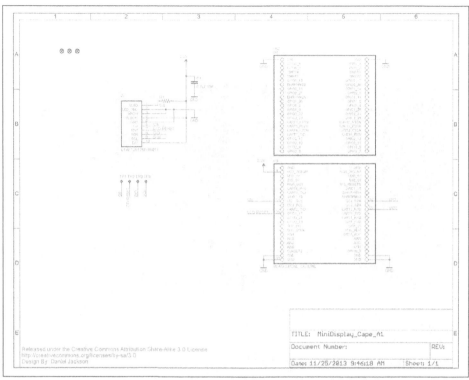

Figure 9-23. Schematic for the MiniDisplay cape

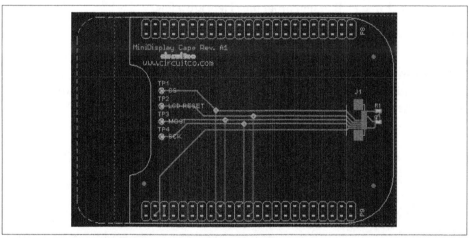

Figure 9-24. PCB for MiniDisplay cape

A good starting point is to take the PCB layout for the MiniDisplay and edit it for your project. The connectors for P8 and P9 are already in place and ready to go.

Eagle PCB is a powerful system with many good tutorials online. The free version runs on Windows, Mac, and Linux, but it has three limitations (*http://bit.ly/1E5Kh3l*):

- The usable board area is limited to 100 x 80 mm (4 x 3.2 inches).
- You can use only two signal layers (Top and Bottom).
- The schematic editor can create only one sheet.

You can install Eagle PCB on your Linux host by using the following command:

```
host$ sudo apt-get install eagle
Reading package lists... Done
Building dependency tree
Reading state information... Done
...
Setting up eagle (6.5.0-1) ...
Processing triggers for libc-bin (2.19-0ubuntu6.4) ...
host$ eagle
```

You'll see the startup screen shown in Figure 9-25.

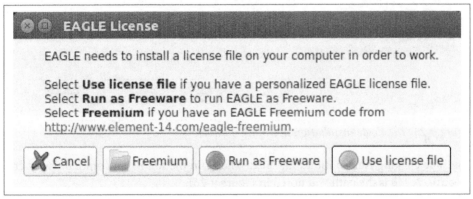

Figure 9-25. Eagle PCB startup screen

Click "Run as Freeware." When my Eagle started, it said it needed to be updated. To update on Linux, follow the link provided by Eagle and download *eagle-lin-7.2.0.run* (or whatever version is current.). Then run the following commands:

```
host$ chmod +x eagle-lin-7.2.0.run
host$ ./eagle-lin-7.2.0.run
```

A series of screens will appear. Click Next. When you see a screen that looks like Figure 9-26, note the Destination Directory.

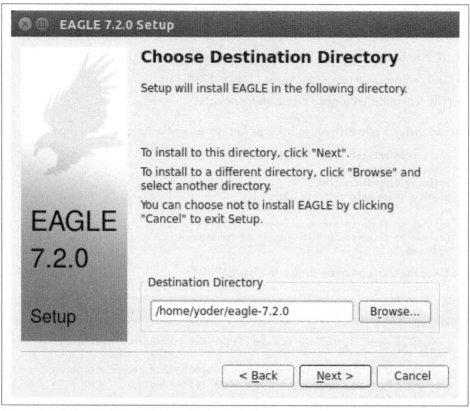

Figure 9-26. The Eagle installation destination directory

Continue clicking Next until it's installed. Then run the following commands (where ~/eagle-7.2.0 is the path you noted in Figure 9-26):

```
host$ cd /usr/bin
host$ sudo rm eagle
host$ sudo ln -s ~/eagle-7.2.0/bin/eagle .
host$ cd
host$ eagle
```

The ls command links eagle in */usr/bin*, so you can run eagle from any directory. After eagle starts, you'll see the start screen shown in Figure 9-27.

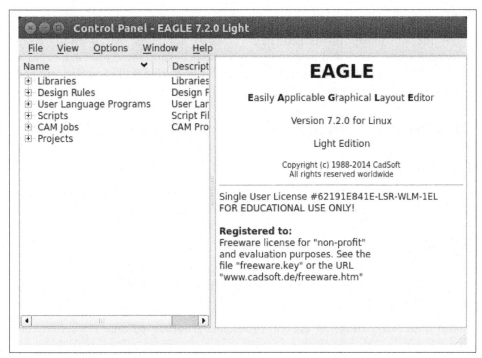

Figure 9-27. The Eagle start screen

Ensure that the correct version number appears.

If you are moving a design from Fritzing to Eagle, see Recipe 9.8 for tips on converting from one to the other.

DesignSpark PCB

The free DesignSpark PCB (*http://bit.ly/19cbwS0*) doesn't have the same limitations as Eagle PCB, but it runs only on Windows. Also, it doesn't seem to have the following of Eagle at this time.

Upverter

In addition to free solutions you run on your desktop, you can also work with a browser-based tool called Upverter (*https://upverter.com/*). With Upverter, you can collaborate easily, editing your designs from anywhere on the Internet. It also provides many conversion options and a PCB fabrication service.

Don't confuse Upverter with Upconverter (Recipe 9.8). Though their names differ by only three letters, they differ greatly in what they do.

Kicad

Unlike the previously mentioned free (no-cost) solutions, Kicad (*http://bit.ly/ 1b2bnBg*) is open source and provides some features beyond those of Fritzing. Notably, CircuitHub (*http://circuithub.com/*) (discussed in Recipe 9.11) provides support for uploading Kicad designs.

9.8 Migrating a Fritzing Schematic to Another Tool

Problem

You created your schematic in Fritzing, but it doesn't integrate with everything you need. How can you move the schematic to another tool?

Solution

Use the Upverter schematic-file-converter (*http://bit.ly/1wXUkdM*) Python script. For example, suppose that you want to convert the Fritzing file for the diagram shown in Figure 9-9. First, install Upverter.

I found it necessary to install `libfreetype6` and `freetype-py` onto my system, but you might not need this first step:

```
host$ sudo apt-get install libfreetype6
Reading package lists... Done
Building dependency tree
Reading state information... Done
libfreetype6 is already the newest version.
0 upgraded, 0 newly installed, 0 to remove and 154 not upgraded.
host$ sudo pip install freetype-py
Downloading/unpacking freetype-py
  Running setup.py egg_info for package freetype-py

Installing collected packages: freetype-py
  Running setup.py install for freetype-py

Successfully installed freetype-py
Cleaning up...
```

 All these commands are being run on the Linux-based host computer, as shown by the `host$` prompt. Log in as a normal user, not `root`.

Now, install the `schematic-file-converter` tool:

```
host$ git clone git@github.com:upverter/schematic-file-converter.git
Cloning into 'schematic-file-converter'...
remote: Counting objects: 22251, done.
remote: Total 22251 (delta 0), reused 0 (delta 0)
Receiving objects: 100% (22251/22251), 39.45 MiB | 7.28 MiB/s, done.
Resolving deltas: 100% (14761/14761), done.
Checking connectivity... done.
Checking out files: 100% (16880/16880), done.
host$ cd schematic-file-converter
host$ sudo python setup.py install
.
.
.
Extracting python_upconvert-0.8.9-py2.7.egg to \
    /usr/local/lib/python2.7/dist-packages
Adding python-upconvert 0.8.9 to easy-install.pth file

Installed /usr/local/lib/python2.7/dist-packages/python_upconvert-0.8.9-py2.7.egg
Processing dependencies for python-upconvert==0.8.9
Finished processing dependencies for python-upconvert==0.8.9
host$ cd ..
host$ python -m upconvert.upconverter -h
usage: upconverter.py [-h] [-i INPUT] [-f TYPE] [-o OUTPUT] [-t TYPE]
                      [-s SYMDIRS [SYMDIRS ...]] [--unsupported]
                      [--raise-errors] [--profile] [-v] [--formats]

optional arguments:
  -h, --help            show this help message and exit
  -i INPUT, --input INPUT
                        read INPUT file in
  -f TYPE, --from TYPE  read input file as TYPE
  -o OUTPUT, --output OUTPUT
                        write OUTPUT file out
  -t TYPE, --to TYPE    write output file as TYPE
  -s SYMDIRS [SYMDIRS ...], --sym-dirs SYMDIRS [SYMDIRS ...]
                        specify SYMDIRS to search for .sym files (for gEDA
                        only)
  --unsupported         run with an unsupported python version
  --raise-errors        show tracebacks for parsing and writing errors
  --profile             collect profiling information
  -v, --version         print version information and quit
  --formats             print supported formats and quit
```

At the time of this writing, Upverter suppports the following file types:

File type	Support
openjson	i/o
kicad	i/o
geda	i/o
eagle	i/o
eaglexml	i/o
fritzing	in only, schematic only
gerber	i/o
specctra	i/o
image	out only
ncdrill	out only
bom (csv)	out only
netlist (csv)	out only

After Upverter is installed, run the file (*quickBot.fzz*) that generated Figure 9-9 through Upverter:

```
host$ python -m upconvert.upconverter -i quickBot.fzz \
-f fritzing -o quickBot-eaglexml.sch -t eaglexml --unsupported
WARNING: RUNNING UNSUPPORTED VERSION OF PYTHON (2.7 > 2.6)
DEBUG:main:parsing quickBot.fzz in format fritzing
host$ ls -l
total 188
-rw-rw-r-- 1 ubuntu ubuntu  63914 Nov 25 19:47 quickBot-eaglexml.sch
-rw-r--r-- 1 ubuntu ubuntu 122193 Nov 25 19:43 quickBot.fzz
drwxrwxr-x 9 ubuntu ubuntu   4096 Nov 25 19:42 schematic-file-converter
```

Figure 9-28 shows the output of the conversion.

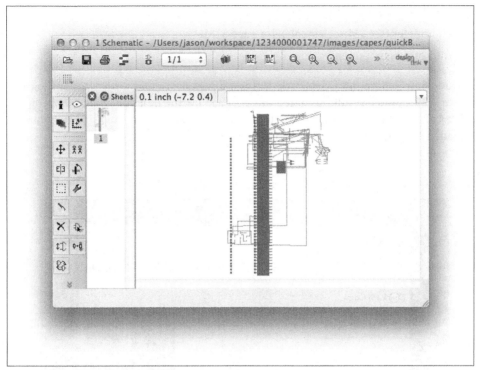

Figure 9-28. Output of Upverter conversion

No one said it would be pretty!

Discussion

I found that Eagle was more generous at reading in the `eaglexml` format than the `eagle` format. This also made it easier to hand-edit any translation issues.

9.9 Producing a Prototype

Problem

You have your PCB all designed. How do you get it made?

Solution

To make this recipe, you will need:

- A completed design (see Recipe 9.7)
- Soldering iron (see "Miscellaneous" on page 318)

- Oscilloscope (see "Miscellaneous" on page 318)
- Multimeter (see "Miscellaneous" on page 318)
- Your other components

Upload your design to OSH Park (*http://oshpark.com*) and order a few boards. Figure 9-29 shows a resulting shared project page for the quickBot cape (*http://bit.ly/ 1MtlzAp*) created in Recipe 9.7. We'll proceed to break down how this design was uploaded and shared to enable ordering fabricated PCBs.

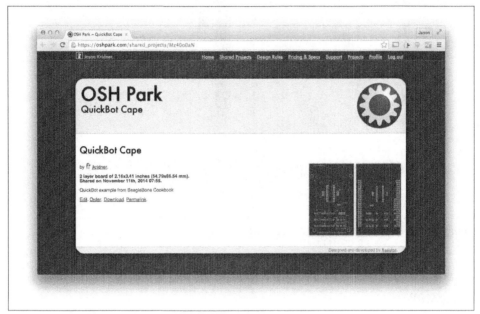

Figure 9-29. The OSH Park QuickBot Cape shared project page

Within Fritzing, click the menu next to "Export for PCB" and choose "Extended Gerber," as shown in Figure 9-30. You'll need to choose a directory in which to save them and then compress them all into a Zip file (*http://bit.ly/1Br5lEh*). The WikiHow article on creating Zip files (*http://bit.ly/1B4GqRU*) might be helpful if you aren't very experienced at making these.

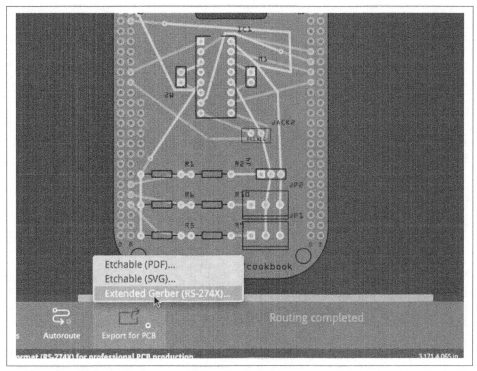

Figure 9-30. Choosing "Extended Gerber" in Fritzing

Things on the OSH Park website (*http://oshpark.com*) are reasonably self-explanatory. You'll need to create an account and upload the Zip file containing the Gerber files (*http://bit.ly/1B4GzEZ*) you created. If you are a cautious person, you might choose to examine the Gerber files with a Gerber file viewer first. The Fritzing fabrication FAQ (*http://bit.ly/18bUgeA*) offers several suggestions, including gerbv (*http://gerbv.source forge.net/*) for Windows and Linux users.

When your upload is complete, you'll be given a quote, shown images for review, and presented with options for accepting and ordering. After you have accepted the design, your list of accepted designs (*https://oshpark.com/users/current*) will also include the option of enabling sharing of your designs so that others can order a PCB, as well. If you are looking to make some money on your design, you'll want to go another route, like the one described in Recipe 9.11. Figure 9-31 shows the resulting PCB that arrives in the mail.

Figure 9-31. QuickBot PCB

Now is a good time to ensure that you have all of your components and a soldering station set up as in Recipe 9.4, as well as an oscilloscope, as used in Recipe 9.6.

When you get your board, it is often informative to "buzz out" a few connections by using a multimeter. If you've never used a multimeter before, the SparkFun (*http://bit.ly/18bUgeA*) or Adafruit (*http://bit.ly/1Br5Xtv*) tutorials might be helpful. Set your meter to continuity testing mode and probe between points where the headers are and where they should be connecting to your components. This would be more difficult and less accurate after you solder down your components, so it is a good idea to keep a bare board around just for this purpose.

You'll also want to examine your board mechanically before soldering parts down. You don't want to waste components on a PCB that might need to be altered or replaced.

When you begin assembling your board, it is advisable to assemble it in functional subsections, if possible, to help narrow down any potential issues. Figure 9-32 shows the motor portion wired up and running the test in Example 9-1.

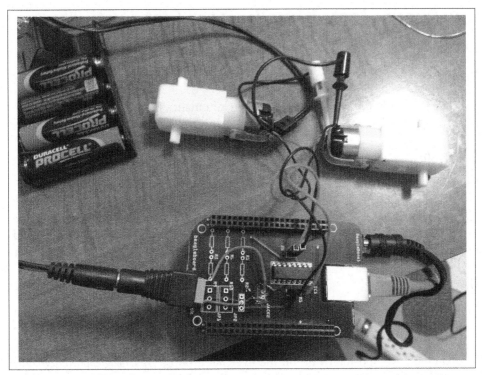

Figure 9-32. QuickBot motors under test

Continue assembling and testing your board until you are happy. If you find issues, you might choose to cut traces and use point-to-point wiring to resolve your issues before placing an order for a new PCB. Better right the second time than the third!

Discussion

There are several solutions for milling a PCB on your own and several alternative online PCB fabrications service vendors. Fritzing even has a fabrication option of its own (*http://fab.fritzing.org*). If you are in Europe, that might be more convenient than OSHPark. I placed an order with Fritzing a few days ahead of placing an order at OSHPark, but the OSHPark order arrived about a week before the order from Fritzing. This is mostly due to the fact I live in the United States.

9.10 Creating Contents for Your Cape Configuration EEPROM

Problem

Your cape is ready to go, and you want it to automatically initialize when the Bone boots up.

Solution

Complete capes have an I²C EEPROM on board that contains configuration information that is read at boot time. Adventures in BeagleBone Cape EEPROMs (*http://bit.ly/1Fb64uF*) gives a helpful description of two methods for programming the EEPROM. How to Roll your own BeagleBone Capes (*http://bit.ly/1E5M7RJ*) is a good four-part series on creating a cape, including how to wire and program the EEPROM.

Discussion

The most important thing to remember when making a cape is that you're working within an ecosystem of Bone developers. Describe the high-level features of your cape and your proposed EEPROM comments to the group mailing list (*http://beagle board.org/discuss*) and follow up on the live chat group (*http://beagleboard.org/chat*) to ensure that you get a response. This is the best way to confirm that your cape is supported by the out-of-box software, making it easier for people buying your cape to get it running.

9.11 Putting Your Cape Design into Production

Problem

You want to share your cape with others. How do you scale up?

Solution

CircuitHub (*https://circuithub.com/*) offers a great tool to get a quick quote on assembled PCBs. To make things simple, I downloaded the CircuitCo MiniDisplay Cape Eagle design materials (*http://bit.ly/1C5uvJc*) and uploaded them to CircuitHub.

After the design is uploaded, you'll need to review the parts to verify that CircuitHub has or can order the right ones. Find the parts in the catalog by changing the text in the search box and clicking the magnifying glass. When you've found a suitable match, select it to confirm its use in your design, as shown in Figure 9-33.

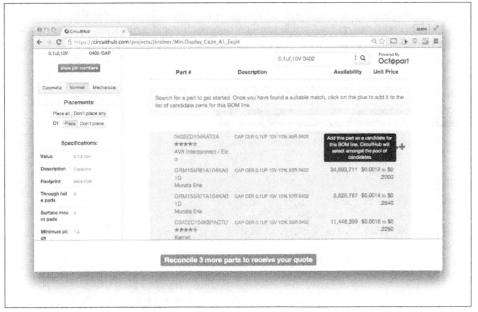

Figure 9-33. CircuitHub part matching

When you've selected all of your parts, a quote tool appears at the bottom of the page, as shown in Figure 9-34.

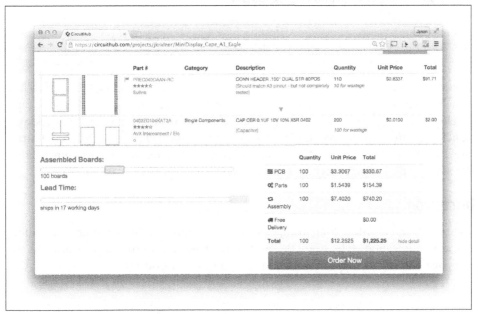

Figure 9-34. CircuitHub quote generation

Checking out the pricing on the MiniDisplay Cape (without including the LCD itself) in Table 9-1, you can get a quick idea of how increased volume can dramatically impact the per-unit costs.

Table 9-1. CircuitHub price examples (all prices USD)

Quantity	1	10	100	1000	10,000
PCB	$208.68	$21.75	$3.30	$0.98	$0.90
Parts	$11.56	$2.55	$1.54	$1.01	$0.92
Assembly	$249.84	$30.69	$7.40	$2.79	$2.32
Per unit	$470.09	$54.99	$12.25	$4.79	$4.16
Total	$470.09	$550.00	$1,225.25	$4,796.00	$41,665.79

Checking the Crystalfontz web page for the LCD (*http://bit.ly/1GF6xqE*), you can find the prices for the LCDs as well, as shown in Table 9-2.

Table 9-2. LCD pricing (USD)

Quantity	1	10	100	1000	10,000
Per unit	$12.12	$7.30	$3.86	$2.84	$2.84
Total	$12.12	$73.00	$386.00	$2,840.00	$28,400.00

To enable more cape developers to launch their designs to the market, CircuitHub has launched a group buy campaign site (*http://campaign.circuithub.com*). You, as a cape developer, can choose how much markup you need to be paid for your work and launch the campaign to the public. Money is only collected if and when the desired target quantity is reached, so there's no risk that the boards will cost too much to be affordable. This is a great way to cost-effectively launch your boards to market!

Discussion

There's no real substitute for getting to know your contract manufacturer, its capabilities, communication style, strengths, and weaknesses. Look around your town to see if anyone is doing this type of work and see if they'll give you a tour.

Don't confuse CircuitHub and CircuitCo. CircuitCo is the official contract manufacturer of BeagleBoard.org and not the same company as CircuitHub, the online contract manufacturing service. CircuitCo would be an excellent choice for you to consider to perform your contract manufacturing, but it doesn't offer an online quote service at this point, so it isn't as easy to include details on how to engage with it in this book.

Parts and Suppliers

Parts

The following tables list where you can find the parts used in this book. We have listed only one or two sources here, but you can often find a given part in many places.

Table A-1. United States suppliers

Supplier	Website	Notes
Adafruit	*http://www.adafruit.com*	Good for modules and parts
Amazon	*http://www.amazon.com/*	Carries everything
Digikey	*http://www.digikey.com/*	Wide range of components
MakerShed	*http://www.makershed.com/*	Good for modules, kits, and tools
RadioShack	*http://www.radioshack.com/*	Walk-in stores
SeeedStudio	*http://www.seeedstudio.com/depot/*	Low-cost modules
SparkFun	*http://www.sparkfun.com*	Good for modules and parts

Table A-2. Other suppliers

Supplier	Website	Notes
Element14	*http://element14.com/BeagleBone*	World-wide BeagleBoard.org-compliant clone of BeagleBone Black, carries many accessories

Prototyping Equipment

Many of the hardware projects in this book use jumper wires and a breadboard. We prefer the preformed wires that lie flat on the board. Table A-3 lists places with jumper wires, and Table A-4 shows where you can get breadboards.

Table A-3. Jumper wires

Supplier	Website
Amazon	*http://www.amazon.com/Elenco-Piece-Pre-formed-Jumper-Wire/dp/B0002H7AIG*
Digikey	*http://www.digikey.com/product-detail/en/TW-E012-000/438-1049-ND/643115*
RadioShack	*http://www.radioshack.com/solderless-breadboard-jumper-wire-kit/2760173.html#.VG5i1PnF8fA*
SparkFun	*https://www.sparkfun.com/products/124*

Table A-4. Breadboards

Supplier	Website
Amazon	*http://www.amazon.com/s/ref=nb_sb_noss_1?url=search-alias%3Dtoys-and-games&field-keywords=breadboards&sprefix=breadboards%2Ctoys-and-games*
Digikey	*http://www.digikey.com/product-search/en/prototyping-products/solderless-breadboards/2359510?k=breadboard*
RadioShack	*http://www.radioshack.com/search?q=breadboard*
SparkFun	*https://www.sparkfun.com/search/results?term=breadboard*
CircuitCo	*http://elinux.org/CircuitCo:BeagleBone_Breadboard*

If you want something more permanent, try Adafruit's Perma-Proto Breadboard (*https://www.adafruit.com/product/1609*), laid out like a breadboard.

Resistors

We use 220 Ω, 1 kΩ, 4.7 kΩ, 10 kΩ, 20 kΩ, and 22 kΩ resistors in this book. All are 0.25 W. The easiest way to get all these, and many more, is to order SparkFun's Resistor Kit (*http://bit.ly/1EXREh8*). It's a great way to be ready for future projects, because it has 500 resistors. RadioShack's 500-piece Resistor Assortment (*http://shack.net/1B4Io4V*) is a bit more expensive, but it has a wider variety of resistors.

If you don't need an entire kit of resistors, you can order a la carte from a number of places. RadioShack has 5-packs (*http://shack.net/1E5NoIC*), and DigiKey has more than a quarter million through-hole resistors (*http://bit.ly/1C6WQjZ*) at good prices, but make sure you are ordering the right one.

You can find the 10 kΩ trimpot (or variable resistor) at SparkFun (*http://bit.ly/18ACvpm*), Adafruit (*http://bit.ly/1NKg1Tv*), or RadioShack (*http://shack.net/1Ag286e*).

Flex resistors (sometimes called *flex sensors* or *bend sensors*) are available at SparkFun (*http://bit.ly/1Br7HD2*) and Adafruit (*http://bit.ly/1HCGoql*).

Transistors and Diodes

The 2N3904 (*http://bit.ly/1B4J8H4*) is a common NPN transistor that you can get almost anywhere. Even Amazon (*http://amzn.to/1AjvcsD*) has it. Adafruit (*http://bit.ly/1b2dgxT*) has a nice 10-pack. SparkFun (*http://bit.ly/1GrZj5P*) lets you buy them one at a time. DigiKey (*http://bit.ly/1GF8H9K*) will gladly sell you 100,000.

The 1N4001 (*http://bit.ly/1EbRzF6*) is a popular 1A diode. Buy one at SparkFun (*http://bit.ly/1Ajw54G*), 10 at Adafruit (*http://bit.ly/1Gs05zP*), 25 at RadioShack (*http://shack.net/1E5OTXi*), or 40,000 at DigiKey (*http://bit.ly/18ADlT2*).

Integrated Circuits

The PCA9306 is a small integrated circuit (IC) that converts voltage levels between 3.3 V and 5 V. You can get it cheaply in large quantities from DigiKey (*http://bit.ly/1Fb8REd*), but it's in a very small, hard-to-use, surface-mount package. Instead, you can get it from SparkFun on a Breakout board (*http://bit.ly/19ceTsd*), which plugs into a breadboard.

The L293D is an H-bridge IC (*http://bit.ly/1wujQqk*) with which you can control large loads (such as motors) in both directions. SparkFun (*http://bit.ly/18bXChR*), Adafruit (*http://bit.ly/1xd43Yh*), and DigiKey (*http://bit.ly/18bXKOk*) all have it in a DIP package that easily plugs into a breadboard.

The ULN2003 is a 7 darlington NPN transistor IC array used to drive motors one way. You can get it from DigiKey (*http://bit.ly/1D5UQIB*). A possible substitution is ULN2803 available from SparkFun (*http://bit.ly/1xd4oKy*) and Adafruit (*http://bit.ly/1EXWhaU*).

The TMP102 is an I²C-based digital temperature sensor. You can buy them in bulk from DigiKey (*http://bit.ly/1EA02Vx*), but it's too small for a breadboard. SparkFun (*http://bit.ly/1GFafAE*) sells it on a breakout board that works well with a breadboard.

The DS18B20 is a one-wire digital temperature sensor that looks like a three-terminal transistor. Both SparkFun (*http://bit.ly/1Fba7Hv*) and Adafruit (*http://bit.ly/1EbSYvC*) carry it.

Opto-Electronics

LEDs (*http://bit.ly/1BwZvQj*) are *light-emitting diodes.* LEDs come in a wide range of colors, brightnesses, and styles. You can get a basic red LED at SparkFun (*http://bit.ly/1GFaHPi*), Adafuit (*http://bit.ly/1GFaH1M*), RadioShack (*http://shack.net/1KWVVGE*), and DigiKey (*http://bit.ly/1b2f2PD*).

Many places carry bicolor LED matrices, but be sure to get one with an I²C interface. Adafruit (*http://bit.ly/18AENVn*) is where I got mine.

Capes

There are a number of sources for capes for BeagleBone Black. BeagleBoard.org (*http://bit.ly/1AjlXJ9*) keeps a current list.

Miscellaneous

Here are some things that don't fit in the other categories.

Table A-5. Miscellaneous

3.3 V FTDI cable	SparkFun (*http://bit.ly/1FMeXsG*), Adafruit (*http://bit.ly/18AF1Mm*)
USB WiFi adapter	Adafruit (*http://www.adafruit.com/products/814*)
Female HDMI to male microHDMI adapter	Amazon (*http://amzn.to/1C5BcLp*)
HDMI cable	SparkFun (*https://www.sparkfun.com/products/11572*)
Micro HDMI to HDMI cable	Adafruit (*http://www.adafruit.com/products/1322*)
HDMI to DVI Cable	SparkFun (*https://www.sparkfun.com/products/12612*)
HDMI monitor	Amazon (*http://amzn.to/1B4MABD*)
Powered USB hub	Amazon (*http://amzn.to/1NKm2zB*), Adafruit (*http://www.adafruit.com/products/961*)
Keyboard with USB hub	Amazon (*http://amzn.to/1FbbISX*)
Soldering iron	SparkFun (*http://bit.ly/1FMfUkP*), Adafruit (*http://bit.ly/1EXZ6J1*)
Oscilloscope	Adafruit (*https://www.adafruit.com/products/468*)
Multimeter	SparkFun (*http://bit.ly/1C5BUbu*), Adafruit (*http://bit.ly/1wXX3np*)

PowerSwitch Tail II	SparkFun (*http://bit.ly/1Ag5bLP*), Adafruit (*http://bit.ly/1wXX8aF*)
Servo motor	SparkFun (*http://bit.ly/1C72cvw*), Adafruit (*http://bit.ly/1HCPQdl*)
5 V power supply	SparkFun (*http://bit.ly/1C72q5C*), Adafruit (*http://bit.ly/18c0n2D*)
3 V to 5 V motor	SparkFun (*http://bit.ly/1b2g65Y*), Adafruit (*http://bit.ly/1C72DWF*)
3 V to 5 V bipolar stepper motor	SparkFun (*http://bit.ly/1Bx2hVU*), Adafruit (*http://bit.ly/18c0HhV*)
3 V to 5 V unipolar stepper motor	Adafruit (*http://www.adafruit.com/products/858*)
Pushbutton switch	SparkFun (*http://bit.ly/1AjDf90*), Adafruit (*http://bit.ly/1b2glhw*)
Magnetic reed switch	SparkFun (*https://www.sparkfun.com/products/8642*)
LV-MaxSonar-EZ1 Sonar Range Finder	SparkFun (*http://bit.ly/1C73dDH*), Amazon (*http://amzn.to/1wXXvIP*)
HC-SR04 Ultrsonic Range Sensor	Amazon (*http://amzn.to/1FbcPNa*)
Rotary encoder	SparkFun (*http://bit.ly/1D5ZypK*), Adafruit (*http://bit.ly/1D5ZGp3*)
GPS receiver	SparkFun (*http://bit.ly/1EA2sn0*), Adafruit (*http://bit.ly/1MrS2VV*)
BLE USB dongle	Adafruit (*http://www.adafruit.com/products/1327*)
SensorTag	DigiKey (*http://bit.ly/18AGPVt*), Amazon (*http://amzn.to/1EA2B9U*), TI (*https://store.ti.com/CC2541-SensorTag-Development-Kit-P3192.aspx*)
Syba SD-CM-UAUD USB Stereo Audio Adapter	Amazon (*http://amzn.to/1EA2Gdl*)
Sabrent External Sound Box USB-SBCV	Amazon (*http://amzn.to/1C74kTU*)
Vantec USB External 7.1 Channel Audio Adapter	Amazon (*http://amzn.to/19cinev*)
Nokia 5110 LCD	Adafruit (*http://bit.ly/1Ag6LgG*), SparkFun (*http://bit.ly/19cizdu*)
BeagleBone LCD7	eLinux (*http://elinux.org/CircuitCo:BeagleBone_LCD7#Distributors*)
MiniDisplay Cape	eLinux (*http://elinux.org/CircuitCo:MiniDisplay_Cape*)

Index

Symbols
# character, 28, 131
$ character, 28
120 V devices, 94
5 V devices, 101, 120
^C (Ctrl-C) command, 52, 96, 113, 117, 120, 267
` character (backtick), 230, 240
~ character, 47

A
ABI documentation, 235
accelerometers, 81
Adafruit BBIO library, 175
Adafruit Bicolor 8x8 LED Square Pixel Matrix, 97
Adafruit Neopixel LED strings, 102
adduser, 136
Advanced Linus Sound Architecture (ALSA), 87
advanced operations
  accessing command shell via SSH, 133
  accessing command shell via virtual serial port, 135
  copying files, 168
  editing text file from GNU/Linux command shell, 149
  Ethernet-based Internet connections, 151
  firewalls, 164
  freeing memory space, 169
  GNU/Linux commands, 147
  installing additional Node.js packages, 173
  installing additional packages, 165
  interactions via C, 176
  interactions via Python, 175
  remote control via VNC server, 143
  removing packages, 167
  running standalone, 129
  selecting host computer OS, 132
  sharing host's Internet connection over USB, 160
  using graphical editors, 150
  verifying OS version from the shell, 142
  viewing system messages, 137, 231
  WiFi-based Internet connections, 154
AM335x Technical Reference Manual, 251
analog voltage sensors, 59
analog-to-digital converter (ADC) inputs, 44, 55
analogWrite(), 95, 110
Apache web servers, 217
applications
  running automatically, 29
  running from Cloud9, 27
apt-get
  installing packages with, 165
  removing packages installed with, 167
Arduino, 221
ARM processor, 245
attachInterrupt(), 49, 50, 65
audio files
  HDMI audio, 132
  recording, 86
  text-to-speech programs, 106
autoremove, 168
Autoroute, 291
autorun folder, 29

B

b.analogWrite(), 95, 110
b.attachInterrupt(), 49, 50, 65
b.digitalRead(), 52
b.pinMode(), 49, 50, 53
backports, 166
backtick (`) character, 230, 240
backups, onboard flash, 39, 49
bash command window, 27, 47
basic operations
    backups, 39, 49
    BeagleBone/BeagleBoard selection, 1
    BoneScript API tutorials, 18
    editing code with Cloud9, 25
    getting started, 10
    OS updates, 17, 30
    running applications automatically, 29
    running JavaScript applications from
        Cloud9, 27
    running latest OS version, 33
    shut-down, 16
    updating onboard flash, 41
    wiring a breadboard, 21
BBIO library, 175
beagleboard.org web page, 15
BeagleBoards
    BeagleBoard-X15, 9
    BeagleBoard-xM, 3
    selecting, 1
BeagleBones
    BeagleBone White, 5
    benefits of, vii, 7
    development ecosystem, 310
    selecting, 1
binary numbers, 251
bipolar stepper motors, 109, 122
blinked.js, 25
Bluetooth Low Energy (BLE), 81
BONE101 GitHub page, 20
BoneScript
    API examples page, 19
    displaying GPIO pin values via, 193
    input/output (I/O) with, 246
    interactive guide, 15
    running applications automatically, 29
    tutorials for, 18
bootable partitions, 36
breadboards
    sensor connection through, 46

suppliers for, 316
vs. protoboards, 280
wiring, 21
bulk email, 201
buses, 23

C

C language
    benefits of interfacing via, 176
    mapping GPIO registers with mmap() and,
        253
    real time I/O with, 248
cape headers
    P8/P9 connection options, 45
    P8/P9 diagram, 44
    stacking vs. non-stacking, 275
capes
    audio, 86
    basics of, 269
    connecting multiple, 275
    design verification, 285
    EEPROM, 275, 310
    LCD display, 128 x 128-pixel, 271
    LCD display, seven-inch, 269
    printed circuit board layout, 291
    protoboards for, 280
    prototype production, 305
    prototype schematics, 281
    scaling up production, 310
    sources of, 318
cd (change directory) command, 28, 47, 169
chmod (change mode) command, 28, 49
Chrome, 12
Chromium web browser, 170
circuit schematics, 281, 302
CircuitCo, 313
CircuitHub, 310
Cloud9
    editing code using, 25
    root user, 27
    running BoneScript applications from, 29
    running JavaScript applications from, 27, 47
    themes, 16
    web page for, 16
    website for, 26
code
    editing with Cloud9, 25
    entering and running on sensors, 46
    obtaining and using examples, xi

---

322 | Index

eMMC memory (see onboard flash)
environmental sensors
    temperature, 74, 78
    TI SensorTag, 81, 216
eQEP2 encoder, 65
exec(), 107
external power supplies, 126, 154

F
fade(), 96
files
    accessing from host computer, 181
    audio, 86, 106, 132
    converting schematics, 302
    copying, 168
    deleting, 172
    discovering large, 170
    editing text, 149
    listing, 182
    patch file application, 241
    patch file creation, 243
    virtual file system for, 231
Firefox, 12
firewalls, 164
flashLED(), 248
flex sensors, 55
flite text-to-speech program, 106
flot plotting package, 195
force-sensitive resistors, 55
Fritzing
    creating circuit schematics with, 281
    exporting schematics from, 302
    installing, xi
    printed circuit board tab in, 291
    prototype production with, 306
FTDI pins, 137

G
global positioning system (GPS), 71, 173
Gmail, 199
GPIO pins
    avoiding burn outs, 24
    controlling via sysfs entries, 233
    displaying status of, 186
    displaying value continuously, 193
    displaying value via jsfiddle, 189
    for sensor connections, 44
    header pin mapping for libsoc, 178
    pull-up and -down resistors, 25, 53

(see also input/output (I/O))
graphical editors, 150
graphics, displaying, 104
gyroscopes, 81

H
H-bridges, 117, 122
halt command, 16
hard real-time systems, 245
hardware
    5 V devices, 101, 120
    building drivers for, 229
    creating illustrations of, xi
    creating prototype schematics, 281
        (see also prototypes)
    high-voltage devices, 94
    radio-controlled, 113
HC-SR04 Ultrasonic Range Sensor, 61
HDMI displays, 129
hex numbers, 251
high-resolution timers, 65
high-voltage devices, 94
hobby servo motors (see servo motors)
host computer
    accessing files from, 181
    accessing graphical desktop from, 143
    editing code using Cloud9, 25
    installing USB drivers for, 13
    Linux OS, 34
    Mac OS, 37
    OS selection, 132
    sharing Internet connection over USB, 160
    Windows OS, 34
hostname, setting, 154
hrtime(), 65
humidity sensor, 81

I
ID.txt, 17
ifconfig command, 152, 155
input mode, 44
input/output (I/O)
    soft vs. hard real-time systems, 245
    via Serial Peripheral Interface (SPI), 273
    with BoneScript, 246
    with C and libsoc, 248
    with C and mmap(), 253
    with devmem2, 251

mmap(), 253
motion sensors, 81
motors
    bipolar stepper control, 122
    DC motor direction control, 117
    DC motor speed control, 114
    servo motor control, 110
    types of, 109
    unipolar stepper control, 126
        (see also input/output (I/O))
mount command, 169
mouse, 129, 271
move(), 113
multimeters, 308

N

nano editor, 149
National Marine Electronics Association
    (NMEA), 72, 173
Neopixel LED strings, 102
networking
    accessing files on host computer, 181
    displaying GPIO pin status, 186
    displaying GPIO value continuously, 193
    displaying GPIO value via jsfiddle, 189
    displaying weather conditions, 202
    graphical programing via Node-RED, 207
    interaction over serial connection, 221
    interaction via web browsers, 184
    Internet connections, 181
    plotting live data, 195
    sending email, 199
    sending SMS messages, 201
    sending/receiving Twitter posts, 204
    serving web pages, 182, 217
Node Package Manager, 173
Node-RED, 207
Node.js web server, 182
Nokia 5510 LCD display, 104
non-stacking headers, 275
normal user, 28, 133

O

on/off sensors, 50
onboard flash
    backing up, 39, 49
    copying files to/from, 168
    extracting, 39
    freeing space on, 169

programming, 33
    updating, 41
online dictionary, installing, 165
onReadByte(), 77
onSerial function, 73
operating system (OS)
    Linux, 34
    Mac, 37
    running the latest, 33
    selecting for host computer, 132
    updating, 17, 30, 38
    verifying from the shell, 142
    Windows, 34
opto-electronics, suppliers for, 318
OSH Park, 306
outputs
    controlling 5 V devices, 101
    controlling high-voltage devices, 94
    DVI-D, 130
    fading external LEDs, 95
    LED matrix, 97
    microHDMI, 129
    NeoPixel LED strings, 102
    Nokia 5510 LCD display, 104
    toggling external LEDs, 92
    toggling onboard LEDs, 90

P

P8/P9 cape headers
    diagram of, 44
    sensor connection options, 45
partitions, 36
parts, sources for, 315
passive on/off sensors, 50
passwords
    changing, 133, 153
    setting, 29
patches, applying, 241
PCA9306 level translator, 101
ping(), 65
pingEnd(), 65
pinMode(), 49, 50, 53
plotting live data, 195
ports
    defaults, 219
    determining current, 219
    port forwarding, 163
position sensors, 55, 65
power supplies, external, 126, 154

pressure sensors, 81
printed-circuit layout, 281
printStatus(), 61
process.on(), 113
programmable real-time units (PRUs), 104,
    245, 261
programs
    installing, 165
    removing, 167
protoboards, 280
prototypes
    equipment for, 316
    production of, 305, 310
    schematics for, 281
    soldering protoboards, 280
PRU Speak, 261
pulse width modulation (PWM), 95, 110, 116
pulse width sensors, 61
pushbutton sensors, 50
pushbutton.js file, 48, 52
Python, 27, 175

Q
quadrature encoder, 65

R
R/C motors (see servo motors)
radio-controlled toys, 113
rails, 23
readRange(), 61
real-time systems, 245
recording audio files, 86
Remmina Remote Desktop Client, 143
remote procedure call (RPC), 191
remove option, 167
resistor divider circuits, 55
resistors
    suppliers for, 316
    toggling external LEDs with, 92
RJ45 connector, 151, 160
root user, 27, 28, 133
rotary encoders, 65

S
schematic-file-converters, 302
scp (secure copy) command, 49
SD cards
    booting from, 33

copying files to/from, 168
formatting, 36
freeing space on, 169
selecting, 38
security issues
    email, 201
    passwords, 29, 133, 153
sensors
    analog or variable voltage, 59
    Bluetooth 4.0 interface for, 81
    connection options, 45
    entering and running code, 46
    interfaces for, 43
    passive on/off, 50
    rotation, 65
    smart, 71
    temperature, 74, 78
    variable pulse width, 61
    variable resistance, 55
SensorTag, 81, 216
Serial Peripheral Interface (SPI), 273
serial ports
    interaction via, 221
    UARTs, 71
    virtual, 135
series resistors, 92
servo motors
    controlling, 110
    motions available, 109
setInterval(), 61, 65, 96, 113
shut down, 16
SMS messages, 201
Socket.IO, 20
soft real-time systems, 245
software
    installing additional Node.js packages, 173
    installing packages, 165
    OS updates, 17, 30, 38
    removing packages, 167
    removing preinstalled, 169
    updating Linux kernel, 227
soldering
    components to protoboards, 280
    prototypes, 305
solenoids, 114
    (see also DC motors)
speakForSelf(), 107
speech, 106
SSH (Secure Shell)

configuration file, 134
connecting via, 133
installing, 49
sshfs command, 181
stacking headers, 275
START.htm, 12, 13
stepper motors
controlling bipolar, 122
controlling unipolar, 126
motions available, 109, 124
types of, 109
streaming data, 71
superusers, 29, 133
suppliers, websites for, 315
sweep(), 113
sysfs virtual file system
controlling GPIO pins with, 233
controlling LEDs with, 231
system messages, viewing, 137, 231

T
tab character, 230
tail command, 231
temperature sensors
Dallas Semiconductor DS18B20, 78
Sensor Tag, 81
SensorTag, 216
TMP102 sensor, 74
text files, editing, 149
text messages, 201
text, displaying, 104
text-to-speech programs, 106
TI SensorTag, 81, 216
TMP102 sensor, 74
.toFixed(), 58
transistors, suppliers for, 317
Trigger input, 61
trimpots, 56
Tukaani Project, 38
tweets, sending/receiving, 204
Twilio, 201
Twitter posts, 204

U
UARTs (serial ports), 71, 221
ULN2003 Darlington Transistor Array IC, 126
unipolar stepper motors, 109, 126
Upconverter, 302
Upverter, 301
USB 2.0 host port, 43
USB audio adapters, 86
USB devices, 46, 155
USB drivers, installing, 13
USB wireless adapters, 154
USER LEDs, 11, 90
usleep() command, 257

V
variable pulse width sensors, 61
variable resistance sensors
reading, 55
three terminal, 56
two terminal, 58
variable voltage sensors, 59
Virtual Network Computing (VNC) server, 143
virtual serial port, 135
voltage sensors, 59

W
weather APIs, 202
web browsers
Chrome Chromium, 170
selecting, 12
web servers
creating your own, 184
using BeagleBone as, 182, 217
Wheezy, 166
wicd-curses, 155
WiFi adapters, 154
wireless networks, connecting to, 154
Workspace file tree, 25

X
Xenomai, 257
xz utility, 38

About the Authors

Prof. Mark A. Yoder is a professor of Electrical and Computer Engineering (ECE) at Rose-Hulman Institute of Technology in Terre Haute, Indiana. In January 2012, he was named the first Lawrence J. Giacoletto Chair in ECE. He received the school's Board of Trustees Outstanding Scholar Award in 2003. Dr. Yoder likes teaching Embedded Linux and Digital Signal Processing (DSP). He is coauthor of two award-winning texts, *Signal Processing First* and *DSP First: A Multimedia Approach*, both with Jim McClellan and Ron Schafer. Mark and his wife, Sarah, have three boys and seven girls ranging in age from 12 to 31 years old.

Jason Kridner is the cofounder of the BeagleBoard.org Foundation, a United States-based 501(c) non-profit corporation existing to provide education in and promotion of the design and use of open source software and hardware in embedded computing. A veteran of Texas Instruments and the semiconductor industry with more than 20 years' experience, Kridner has deep insights into the future of electronics, pioneering both Texas Instruments' and the semiconductor industry's open source efforts and engagements with open hardware. In his free time, Kridner uses BeagleBone Black to explore his creativity with creations like the StacheCam, which uses a webcam and computer vision to detect faces and superimpose fancy mustaches.

Colophon

The animal on the cover of *BeagleBone Cookbook* is a *Harrier (canis lupus familiaris)*. The Harrier is a medium, hound-bred dog derived from other hound breeds. The Harrier was developed in the UK in the late 13th century and bred for hunting purposes. They are closely related to the Beagle breed, but are not as popular as domesticated pets. The Harrier breed is often referred to as "Beagles on steroids."

Full-grown Harriers can average in size between 19 and 21 inches tall and weigh between 45 and 50 pounds. The breed's coat can consist of two to three colors that might include a combination of white, tan, black, or red. They strongly resemble English Foxhounds but are smaller in size. Although the Harrier is a short-haired breed, they still need to be brushed at least once a week to remove dead hair. They are high-energy dogs and having a large, outdoor area for exercise is preferable, but be sure it is fenced in, as they will take off after smaller animals.

The Harrier is a good breed for household pets. This is most likely due to their friendly demeanor and ability to get along with people of all ages, as well as other dogs. They prefer to be in the company of others, but have the hunter's instinct to prey on smaller animals with unfamiliar scents. Harriers should be exposed to other animals early in their lives if there is anticipation of having multiple, non-canine pets in the home. The breed is also known to be vocal with baying, so be prepared to be alerted at anything that may pique the interest of the Harrier.

Many of the animals on O'Reilly covers are endangered; all of them are important to the world. To learn more about how you can help, go to animals.oreilly.com.

The cover image is from Lydekker's Royal Natural History. The cover fonts are URW Typewriter and Guardian Sans. The text font is Adobe Minion Pro; the heading font is Adobe Myriad Condensed; and the code font is Dalton Maag's Ubuntu Mono.

Get even more for your money.

Join the O'Reilly Community, and register the O'Reilly books you own. It's free, and you'll get:

- $4.99 ebook upgrade offer
- 40% upgrade offer on O'Reilly print books
- Membership discounts on books and events
- Free lifetime updates to ebooks and videos
- Multiple ebook formats, DRM FREE
- Participation in the O'Reilly community
- Newsletters
- Account management
- 100% Satisfaction Guarantee

Signing up is easy:

1. Go to: oreilly.com/go/register
2. Create an O'Reilly login.
3. Provide your address.
4. Register your books.

Note: English-language books only

To order books online:
oreilly.com/store

For questions about products or an order:
orders@oreilly.com

To sign up to get topic-specific email announcements and/or news about upcoming books, conferences, special offers, and new technologies:
elists@oreilly.com

For technical questions about book content:
booktech@oreilly.com

To submit new book proposals to our editors:
proposals@oreilly.com

O'Reilly books are available in multiple DRM-free ebook formats. For more information:
oreilly.com/ebooks

O'REILLY®

CPSIA information can be obtained at www.ICGtesting.com
Printed in the USA
LVOW03s0455280415

436356LV00031B/729/P